CHARACTERIZATIONS AND ANALYSIS OF BLOCK DESIGNS

A.K. NIGAM
Institute of Applied Statistics
and Development Studies
Lucknow, India

P.D. PURI
Sebha University
Sebha, Libya

V.K. GUPTA
Indian Agricultural Statistics Research Institute
New Delhi, India

JOHN WILEY & SONS
New York Chichester Brisbane Toronto Singapore

First Published in 1988 by
WILEY EASTERN LIMITED
4835/24 Ansari Road, Daryaganj
New Delhi 110 002, India

Distributors:

Australia and New Zealand:
Jacaranda-Wiley Ltd., Jacaranda Press,
JOHN WILEY & SONS, INC.
GPO Box 859, Brisbane, Queensland 4001, Australia

Canada:
JOHN WILEY & SONS CANADA LIMITED
22 Worcester Road, Rexdale, Ontario, Canada

Europe and Africa:
JOHN WILEY & SONS LIMITED
Baffins Lane, Chichester, West Sussex, England

South East Asia:
JOHN WILEY & SONS, INC.
05–05 Block B, Union Industrial Building
37 Jalan Pemimpin, Singapore 2057

Africa and South Asia:
WILEY EASTERN LIMITED
4835/24 Ansari Road, Daryaganj
New Delhi 110 002, India

North and South America and rest of the world:
JOHN WILEY & SONS, INC.
605 Third Avenue, New York, N.Y. 10158, USA

Library of Congress Cataloging in Publication Data

Nigam, A.K.
Characterizations and analysis of block designs.

ISBN 0-470-21051-6 John Wiley & Sons, Inc.
ISBN 81-224-0057-4 Wiley Eastern Limited

Printed in India at Prabhat Press, Meerut.

Preface

The purpose of this book is to present the fundamental concepts in the analysis and characterization of block designs. Essentially it incorporates the important developments in the subject that have taken place in the last two decades. The motivation to write this book came from while teaching the subject of experimental designs to the M.Sc. and Ph. D. students of Agricultural Statistics of the Indian Agricultural Research Institute and Haryana Agricultural University. The temptation to write this book followed in particular while teaching the courses on factorial experiments and incomplete block designs to the Ph.D. students of Agricultural Statistics and in fact Chapter 5 on block designs with factorial structure is the backbone of the book.

The book is written for anyone who is engaged in learning the subject on incomplete block designs or is engaged in teaching the subject. It is also meant for research students/scientists who have interest in carrying out further research on the subject. Persons who are engaged in applied research in Universities and state Agricultural Departments, principally from such departments as statistics, agriculture, industry and engineering, should find the method of analysis that are presented to be relevant and useful in their research. As a text book on the subject of incomplete block designs, the contents could serve quite neatly as about two-semester course in most applied curricula.

Since this is the first addition, it has been necessary to exercise considerable selectivity in the choice of topics covered and the manner in which these are covered. Hence no claim is made that the coverage and presentation is exhaustive in either scope or depth. However, we feel that the reader who works through the chapters of this book will achieve a high level working knowledge of the subject and will be tempted to pursue the quest for further research on this subject because there are still many unsolved problems on this subject.

The mathematical prerequisites have been kept at a minimum. Some background knowledge of the matrix algebra and theory of linear estimation is helpful although the results used in the book have been given in Chapter 1 as refresher to readers.

The first Chapter introduces the subject and gives some useful results on linear estimation. These results are used in the subsequent Chapters. Chapter 2 introduces the original block design problem and gives a general procedure of analysis of block designs. It also introduces the famous $\mathbf{\Omega}^{-1}$ matrix of Tocher and gives an iterative procedure of obtaining the inverse of $\mathbf{\Omega}^{-1}$ matrix. This is a useful result because $\mathbf{\Omega}^{-1}$ matrix is a pivot alongwhich revolves the entire analysis of block designs. The Chapter also gives some characterizations of designs based on $\mathbf{\Omega}^{-1}$ matrix.

In Chapter 3 the analysis of balanced and partially balanced designs is dealt with. The useful concepts of variance balanced and efficiency balanced designs are defined and some results are proved on the efficiency balanced and partially efficiency balanced designs. This is a new concept and it brings about quite a simplification in the analysis. The reader is advised to have a grip on this topic because it is this topic which will be used frequently throughout the book.

If designs are obtained by taking Kronecker product of some basic designs like balanced or partially balanced designs then the resulting design has a very simplified analysis in terms of the basic designs. Kronecker product designs form the subject matter of Chapter 4. This Chapter must be read in conjunction with Chapter 5 which is the most important and useful Chapter of the book. The analysis of block designs as factorials and vice versa is the main theme of this Chapter. The concept of factorially balanced and some related results have been described in this Chapter.

Cyclic designs have formed Chapter 6 of the book. Resolvable cyclic designs, generalized cyclic designs and factorial structure in generalized cyclic designs are some of the topics discussed in this Chapter.

Chapter 7 presents many other useful block designs. Dual designs, linked block designs, supplemented block designs, designs with merged treatments, designs with missing blocks, α–designs, nearly balanced designs is the theme of this Chapter. This Chapter also includes the topic on trend free and nearly trend free block designs of Prof. R.A. Bradley and his coworkers. Incomplete block designs for bio-assays also have been considered in this Chapter.

The last chapter discusses some important properties of block designs. These are optimality, robustness and resistant designs. The Grubb's estimator is also described in this Chapter.

We are extremely grateful to many friends for their help in compiling the material for this work. In particular, we are indebted to Prof. Ralph A. Bradley, (University of Georgia, Athens), with whom we had the pleasure of exchanging the ideas regarding the coverage of the book. The topics on 'trend free block designs' and 'Grubb's estimators' are a consequence of his suggestion to devote space to such topics in a modern book like this. As a matter of fact, Prof. Bradley has been a source of inspiration and constant encouragement in preparation of the manuscript of this book. We are very much indebted also to Dr. Rahul Mukerjee (Indian Statistical Institute, Calcutta), and Dr. A. Dey (Indian Agricultural Statistics Research

Institute, New Delhi) who reviewed the initial drafts of this book; their detailed and thoughtful comments on the style and content were instrumental in its organisation. However, despite these associations, whatever errors appear in this book are ours and ours alone. We wish to thank the staff at Wiley Eastern Limited for their encouragements in completing this work. Thanks are also due to the Indian Council of Agricultural Research for granting us permission to get the book published.

A.K. NIGAM
P.D. PURI
V. K. GUPTA

Contents

1. Introduction

1.1 BACKGROUND

The variation associated with the observations collected through an experimental design can be split up into two major components: (i) assignable or controllable variation arising due to a deliberate attempt on the part of the experimenter either to cause or isolate some variation and (ii) variation arising out of the uncontrollable factors, also termed as experimental error. Any attempt to reduce the experimental error results in more precise inference on the experimental treatments. The main thrust of the activity in construction of experimental designs has therefore been at evolving suitable techniques for the reduction of experimental error.

The simplest experimental design is the completely randomized design in which the entire experimental material is assumed to be homogeneous and the assignable part of variation is only due to the application of treatments. All other variations taken together constitute the experimental error. When the experimental material is heterogeneous, it is profitable to reduce the experimental error by forming blocks or rows and columns, etc. The variation arising due to these components is put into the category of assignable part of total variability. This process is known as local control. The designs adopted here are designs for one way elimination of heterogeneity or block designs, designs for elimination of heterogeneity in two directions or row and column designs, multidimensional designs, etc.

Besides local control, the designing of experiments is guided by two more principles. These are randomization and replication. The process of randomization ensures independence of the observations which is necessary for drawing valid inferences by applying suitable statistical techniques. It helps in making an objective comparison among the treatments. On the other hand replication is required for obtaining a valid estimate of experimental error and to some extent for increasing the precision of estimation of treatment contrasts. Kempthorne's (1952) book 'The Design and Analysis of Experiments' gives an excellent account of these three basic principles of experimental design.

The present book is devoted to the study of designs useful in experiments involving one way elimination of heterogeneity. These designs are also termed as block designs. The simplest block design is the randomized complete block design. This design is not only combinatorially simple but is most convenient to analyse. This is an orthogonal design and as such it possesses the optimal property. The design is balanced as well.

Two main problems associated with the theory of experimental designs are (a) the construction and combinatorial problems, and (b) the analysis of the data gathered from the design. The type and nature of analysis as also the properties a design possesses depends heavily on the way the design is constructed. Although analysis of the observations collected through any general unstructured design can be handled on computer, yet in most cases it may not have any appeal for the experimenter as he is completely in dark about the properties of the design until the analysis is complete. On the contrary, if the design is structured, it is possible to have a first hand knowledge of the merits or otherwise of the design. Through appropriate characterizations and classifications of structured designs it may be possible to assess in advance whether or not the contrasts of interest can be estimated with a desired level of efficiency. Besides ease in analysis, such structured designs and their characterization therefore carry a much better appeal and are likely to be of immense use to the experimenter. The main theme of this book is to study the characterizations and analysis of such block designs. Emphasis throughout is on analysis and properties of block designs instead of the combinatorial and construction problems associated with such designs.

Block designs are used mostly in agricultural experiments. In fact the entire development of the subject on block designs originated from the agricultural experiments conducted at the Rothamsted Experimental Station, England. Recently, block designs have found an interesting application in sample selection in survey sampling. The earliest work on establishing a linkage between experimental designs and survey sampling is by Chakrabarti (1963). Among other workers are Singh and Raghavarao (1975), Foody and Hedayat (1977), Wynn (1977), and Das (1980). Recently, Gupta, Nigam and Kumar (1982), Nigam and Gupta (1984), Nigam, Kumar and Gupta (1984), and Kumar, Gupta and Nigam (1985) have demonstrated the usefulness of block designs in unequal probability without replacement sampling schemes.

Before we go on to the subject matter of block designs, we present here some results on linear estimation which shall be used throughout the text.

1.2 SOME RESULTS ON LINEAR ESTIMATION

Let $\mathbf{y} = (y_1,\ldots,y_n)'$ be an n-component random vector of uncorrelated observations and $\mathbf{X}$ be a real matrix of order $n \times k$ $(k < n)$ of known coefficients x_{ij}, $i = 1,\ldots, n$; $j = 1,\ldots,k$. We say that the distribution of $\mathbf{y}$ satisfies a linear model if

$$E(\mathbf{y}) = \mathbf{X\theta},\ D(\mathbf{y}) = \sigma^2\mathbf{I}_n,$$

where σ^2 and $\boldsymbol{\theta} = (\theta_1, \ldots, \theta_k)'$, a k-component vector, are the parameters unknown partly or completely, and $\mathbf{I}_n$ is an identity matrix of order n. The model is linear because we envisage $E(\mathbf{y})$ as being a linear combination of the unknown parameters. It is convenient to write

$$\mathbf{y} = \mathbf{X\theta} + \mathbf{e}, \qquad E(\mathbf{e}) = \mathbf{0}, \; D(\mathbf{e}) = \sigma^2 \mathbf{I}_n, \tag{1.1}$$

where $\mathbf{e} = (e_1, \ldots, e_n)'$ is an n-component vector of non-observable random variables also termed as error component. $\mathbf{y}$ is also assumed to have a multivariate normal distribution $N_n(\mathbf{X\theta}, \sigma^2\mathbf{I}_n)$.

If all the effects (parameters) in the model are fixed (apart from the error component $\mathbf{e}$ which is always random) the model is called the fixed effects model or Model-I of Eisenhart (1947). Further, if all the effects in the model are random (apart from the general mean μ which is always fixed) the model is called the random effects model or Model-II of Eisenhart (1947).

Another type of models, where some of the effects (other than the general mean μ and error $\mathbf{e}$) are fixed and some are random, are called mixed models. Of course, any model which contains a general mean μ and a random error $\mathbf{e}$ is truly a mixed model, but this description is reserved for models whose other effects are a mixture of fixed and random effects. The model of our concern throughout the book is the fixed effects model.

Using the principle of least squares the unknown parameters can be estimated by minimising the sum of squares

$$\mathbf{e}'\mathbf{e} = (\mathbf{y} - \mathbf{X\theta})' \, (\mathbf{y} - \mathbf{X\theta}).$$

This gives $\mathbf{X}'\mathbf{X}\,\boldsymbol{\theta} = \mathbf{X}'\mathbf{y}$. The minimising value $\boldsymbol{\theta}$ is known as a least square estimate of $\boldsymbol{\theta}$. The equations $\mathbf{X}'\mathbf{X}\,\boldsymbol{\theta} = \mathbf{X}'\mathbf{y}$ are called normal equations.

We now examine conditions under which normal equations provide a solution. We state the following Lemmas without proof (see Searle, 1971; Rao, 1973).

Lemma 1.1: A system of non-homogeneous equations $\mathbf{AZ} = \mathbf{b}$ is consistent if and only if $R(\mathbf{A}) = R(\mathbf{A}:\mathbf{b})$, where $R(\cdot)$ is the rank of a matrix; in other words $\mathbf{b}$ belongs to the column space of matrix $\mathbf{A}$. The matrix $\mathbf{A}$ is of order $m \times n$ $(n < m)$, $\mathbf{Z}$ is an n-component vector and $\mathbf{b}$ is a column vector of order m.

Lemma 1.2: If a system of non-homogeneous equations $\mathbf{AZ} = \mathbf{b}$ is consistent, then $\mathbf{Z} = \mathbf{Gb}$ is a solution of the equations $\mathbf{AZ} = \mathbf{b}$ if and only if $\mathbf{AGA} = \mathbf{A}$. The matrix $\mathbf{G}$, of order $n \times m$ is called a generalized inverse of $\mathbf{A}$. A generalized inverse is also known as a g-inverse. It is represented as $\mathbf{A}^-$, i.e., $\mathbf{A}^-$ is a g-inverse of matrix $\mathbf{A}$. A generalized inverse is not unique.

Lemma 1.3: A general solution of the consistent equations $\mathbf{AZ} = \mathbf{b}$ is

$$\mathbf{Z} = \mathbf{Gb} + (\mathbf{GA}-\mathbf{I})\,\mathbf{x}, \tag{1.2}$$

where $\mathbf{x}$ is an arbitrary vector of order n.

Lemma 1.4: For the consistent equations $\mathbf{AZ} = \mathbf{b}$, all solutions for any specified $\mathbf{G}$ are generated by (1.2).

This result is important because only one generalized inverse $\mathbf{G}$ of $\mathbf{A}$ is required to get all solutions to $\mathbf{AZ} = \mathbf{b}$. There can be no solutions other than those generated from (1.2).

Lemma 1.5: Let $\mathbf{H} = \mathbf{GA}$. Then

i) the matrix $\mathbf{H}$ is idempotent, i.e. $\mathbf{H}^2 = \mathbf{H}$.
ii) $R(\mathbf{H}) = R(\mathbf{A})$.

Lemma 1.6: Let the matrix $\mathbf{B}$ of order n be idempotent of rank $r(\leqslant n)$. Then

i) $R(\mathbf{B}) = \text{tr}(\mathbf{B})$, where tr stands for trace of the matrix.
ii) The eigen values of $\mathbf{B}$ are zero and unity, the number of non-zero eigen values being r.
iii) $\mathbf{I}-\mathbf{B}$ is an idempotent matrix.

In the study of linear models (1.1), the matrix $\mathbf{X'X}$ is in frequent use. Some results on the matrix $\mathbf{X'X}$ are given below:

Lemma 1.7: If $\mathbf{G}$ is a g-inverse of $\mathbf{X'X}$, then

i) $\mathbf{G'}$ is a g-inverse of $\mathbf{X'X}$.
ii) $\mathbf{X\,GX'X} = \mathbf{X}$, i.e., $\mathbf{GX'}$ is a g-inverse of $\mathbf{X}$.
iii) $\mathbf{X\,GX'}$ is symmetric, idempotent and invariant with respect to any choice of a g-inverse.

Corollary 1.1: i) $\mathbf{G'X'}$ is a g-inverse of $\mathbf{X}$.
ii) $\mathbf{XG}$ and $\mathbf{XG'}$ are both g-inverses of $\mathbf{X'}$.
iii) $\mathbf{X\,G'X'} = \mathbf{X\,GX'}$ and $\mathbf{X\,G'X'}$ is symmetric.

We now describe some methods of obtaining a generalized inverse of $\mathbf{X'X}$.

Lemma 1.8: Let $R(\mathbf{X}) = m(< k)$. If there exists a matrix $\mathbf{H}$ of order $(k-m)\times k$ such that $R(\mathbf{X'}:\mathbf{H'}) = k$, then

$$R(\mathbf{X'X} + \mathbf{H'H}) = k \text{ and } \mathbf{G} = (\mathbf{X'X} + \mathbf{H'H})^{-1}$$

is a generalized inverse of $\mathbf{X'X}$.

Lemma 1.9: Let $\lambda_1, \ldots, \lambda_m$ be the non-zero eigen values of $\mathbf{X'X}$ and $\xi_1, \ldots, \xi_m$ be the corresponding eigen vectors. Then

$$\mathbf{X}'\mathbf{X} = \sum_{i=1}^{m} \lambda_i \xi_i \xi_i'.$$

A g-inverse of $\mathbf{X}'\mathbf{X}$ is

$$\mathbf{G} = \sum_{i=1}^{m} \lambda_i^{-1} \xi_i \xi_i'.$$

Lemma 1.10: Let $\mathbf{X}'\mathbf{X}$ be a $k \times k$ matrix of rank m. Then by a suitable interchange of rows and columns of matrix $\mathbf{X}'\mathbf{X}$, we can write

$$\mathbf{A} = \mathbf{R}\mathbf{X}'\mathbf{X}\,\boldsymbol{\zeta} = \begin{bmatrix} \mathbf{B} & \mathbf{C} \\ \mathbf{D} & \mathbf{E} \end{bmatrix},$$

where the sub-matrix $\mathbf{B}$ of order m is a non singular matrix, and $\mathbf{R}$ and $\boldsymbol{\zeta}$ represent the elementary row and column operations respectively. Then

$$\mathbf{G} = \boldsymbol{\zeta} \begin{bmatrix} \mathbf{B}^{-1} & \mathbf{0} \\ \mathbf{0} & \mathbf{0} \end{bmatrix} \mathbf{R}$$

is a g-inverse of $\mathbf{X}'\mathbf{X}$. Here $\mathbf{0}$ is a null matrix.

The subject on block designs studied in this book involves reduced normal equations of the type $\mathbf{A}\boldsymbol{\tau} = \mathbf{Q}$, where matrix $\mathbf{A}$ is of order $v \times v$, $\mathbf{A1} = \mathbf{0}$, and $R(\mathbf{A}) = v-1$. $\mathbf{1}$ represents a column vector with unitary elements. These equations are always consistent. We shall use the following result throughout.

Theorem 1.1: Let $\mathbf{A}$ be real, symmetric matrix of order v satisfying $\mathbf{A1} = \mathbf{0}$ and rank $(v-1)$. Let $q (\neq 0)$ be a real scalar constant and $\mathbf{Z}$ be any arbitrary v-component vector such that $\mathbf{Z}'\mathbf{1} \neq 0$. Then

$$R(\mathbf{A} + q\mathbf{Z}\mathbf{Z}') = v.$$

Proof: Let $\xi_1, \ldots, \xi_{v-1}$ be the eigen vectors of $\mathbf{A}$ corresponding to the non-zero eigen values $\lambda_1, \ldots, \lambda_{v-1}$. Condsider the matrix

$$\mathbf{B} = [\sqrt{\lambda_1}\xi_1 \ldots \sqrt{\lambda_{v-1}}\xi_{v-1} \; \sqrt{q}\mathbf{Z}].$$

Since $\xi_i'\xi_j = 0 \;\forall\; i, j = 1, \ldots, v-1, i \neq j$, and $\xi_i'\mathbf{1} = 0$, $i = 1, \ldots, v-1$, and further more $\mathbf{Z}'\mathbf{1} \neq 0$, it follows that rank of matrix $\mathbf{B}$ is v. Moreover, rank of matrix $\mathbf{BB}'$ is also v. But

$$\mathbf{BB}' = \sum_{i=1}^{v-1} \lambda_i \xi_i \xi_i' + q\mathbf{Z}\mathbf{Z}' = \mathbf{A} + q\mathbf{Z}\mathbf{Z}'$$

Hence the proof.

Corollary 1.2: A generalized inverse of matrix $\mathbf{A}$ is $(\mathbf{A} + q\mathbf{Z}\mathbf{Z}')^{-1}$.

Recall the normal equations $\mathbf{X}'\mathbf{X}\boldsymbol{\theta} = \mathbf{X}'\mathbf{y}$. These equations are consistent and always admit a solution, since $\mathbf{X}'\mathbf{y} \in \eta(\mathbf{X}'\mathbf{X})$, where $\eta(\mathbf{X}'\mathbf{X})$ is the linear manifold generated by the columns of $\mathbf{X}'\mathbf{X}$. Moreover, $\eta(\mathbf{X}'\mathbf{X}) = \eta(\mathbf{X}')$, which implies that if there exists a vector $\mathbf{L}$ such that $\mathbf{X}'\mathbf{y} = \mathbf{X}'\mathbf{L}$, then there exists a vector $\boldsymbol{\lambda}$ such that $\mathbf{X}'\mathbf{y} = \mathbf{X}'\mathbf{X}\boldsymbol{\lambda}$ (see Rao, 1973). The minimum of

$$\mathbf{e}'\mathbf{e} = (\mathbf{y} - \mathbf{X}\boldsymbol{\theta})'\,(\mathbf{y} - \mathbf{X}\boldsymbol{\theta})$$

is attained at $\boldsymbol{\theta} = \hat{\boldsymbol{\theta}}$ and is unique for all solutions $\hat{\boldsymbol{\theta}}$ of $\mathbf{X}'\mathbf{X}\boldsymbol{\theta} = \mathbf{X}'\mathbf{y}$. The minimum value of $\mathbf{e}'\mathbf{e}$ is $(\mathbf{y} - \mathbf{X}\hat{\boldsymbol{\theta}})'\,(\mathbf{y} - \mathbf{X}\hat{\boldsymbol{\theta}})$.

Assume that $R(\mathbf{X}) = R(\mathbf{X}'\mathbf{X}) = m(< k)$. We now consider the estimable parametric functions. The least squares estimator of the parametric function $\mathbf{P}'\boldsymbol{\theta}$ is defined to be $\mathbf{P}'\hat{\boldsymbol{\theta}}$, where $\hat{\boldsymbol{\theta}}$ is any solution of $\mathbf{X}'\mathbf{X}\boldsymbol{\theta} = \mathbf{X}'\mathbf{y}$. $\mathbf{P}'\hat{\boldsymbol{\theta}}$ is linear in observations and unbiased for $\mathbf{P}'\boldsymbol{\theta}$ if and only if $\mathbf{P}'\hat{\boldsymbol{\theta}}$ is unique for all solution $\hat{\boldsymbol{\theta}}$ of $\mathbf{X}'\mathbf{X}\boldsymbol{\theta} = \mathbf{X}'\mathbf{y}$, or equivalently $\mathbf{P} \in \eta(X')$ or $\mathbf{P} \in \eta(\mathbf{X}'\mathbf{X})$, or equivalently there exists a linear function of observations, $\mathbf{L}'\mathbf{y}$, such that $E(\mathbf{L}'\mathbf{y}) = \mathbf{P}'\boldsymbol{\theta}$. Such a function $\mathbf{P}'\boldsymbol{\theta}$ is called estimable. Further, if $\mathbf{P}'\boldsymbol{\theta}$ is estimable then $\mathbf{P}'\hat{\boldsymbol{\theta}}$ has minimum variance in the class of linear unbiased estimators of $\mathbf{P}'\boldsymbol{\theta}$. Also, if $\mathbf{P}'\hat{\boldsymbol{\theta}}$ and $\mathbf{R}'\hat{\boldsymbol{\theta}}$ are the least squares estimators of the estimable functions $\mathbf{P}'\boldsymbol{\theta}$ and $\mathbf{R}'\boldsymbol{\theta}$, then

$$V(\mathbf{P}'\hat{\boldsymbol{\theta}}) = \sigma^2\mathbf{P}'(\mathbf{X}'\mathbf{X})^-\,\mathbf{P},\ \mathrm{Cov}\,(\mathbf{P}'\hat{\boldsymbol{\theta}}, \mathbf{R}'\hat{\boldsymbol{\theta}}) = \sigma^2\mathbf{P}'(\mathbf{X}'\mathbf{X})^-\,\mathbf{R}.$$

The minimum sum of square $R_0^2 = (\mathbf{y} - \mathbf{X}\hat{\boldsymbol{\theta}})'(\mathbf{y} - \mathbf{X}\hat{\boldsymbol{\theta}})$ is helpful in obtaining an unbiased estimate of σ^2. Infact $E(R_0^2) = (n - m)\sigma^2$, where $R(\mathbf{X}) = m$, so that an unbiased estimator of σ^2 is

$$\hat{\sigma}^2 = R_0^2/(n - m).$$

We now state the following results on estimable parametric functions.

Lemma 1.11: (i) The expected value of any observation is estimable.
(ii) Any linear function of estimable parametric functions is estimable.
(iii) The total number of linearly independent estimable parametric functions in model (1.1) is equal to $R(\mathbf{X})$.

We now state some results on testing of hypothesis and analysis of variance. Consider r linearly independent parametric functions $\mathbf{H}'\boldsymbol{\theta}$, individually estimable under the model (1.1). In other words the matrix $\mathbf{H}'$ of order $r \times k$ and row rank of r is such that the rows of $\mathbf{H}'$ belong to the row space of $\mathbf{X}$. The least squares estimators of $\mathbf{H}'\boldsymbol{\theta}$ are

$$\mathbf{Z} = \mathbf{H}'\hat{\boldsymbol{\theta}} = \mathbf{H}'(\mathbf{X}'\mathbf{X})^-\,\mathbf{X}'\mathbf{y} \quad \text{and} \quad R_0^2 = \mathbf{y}'[\mathbf{I} - \mathbf{X}(\mathbf{X}'\mathbf{X})^-\,\mathbf{X}']\mathbf{y}.$$

Then

$$\mathbf{H}'(\mathbf{X}'\mathbf{X})^-\,\mathbf{X}'[\mathbf{I} - \mathbf{X}(\mathbf{X}'\mathbf{X})^-\,\mathbf{X}'] = \mathbf{0},$$

showing that $\mathbf{H}'\hat{\boldsymbol{\theta}}$ and R_0^2 are independently distributed. The distribution of $\mathbf{H}'\hat{\boldsymbol{\theta}}$ is r variate normal with mean $\mathbf{H}'\boldsymbol{\theta}$ and dispersion matrix $\sigma^2\mathbf{H}'(\mathbf{X}'\mathbf{X})^-\mathbf{H} = \sigma^2\,\mathbf{D}$ (say) and that of R_0^2 is $\sigma^2\chi^2_{n-m}$. Also $(\mathbf{H}'\hat{\boldsymbol{\theta}} - H'\boldsymbol{\theta})'\,\mathbf{D}^{-1}(\mathbf{H}'\hat{\boldsymbol{\theta}} - \mathbf{H}'\boldsymbol{\theta})$ follows a $\sigma^2\chi^2_r$ distribution.

Let it be desired to test the null hypothesis

$$H_0 : \mathbf{H}'\boldsymbol{\theta} = \boldsymbol{\theta}_0.$$

Consider the vector of deviations $(\mathbf{Z} - \boldsymbol{\theta}_0)$. Then under H_0

$$E(\mathbf{Z} - \boldsymbol{\theta}_0)'\mathbf{D}^{-1}(\mathbf{Z} - \boldsymbol{\theta}_0)$$
$$= \sigma^2 + \frac{1}{r}(\mathbf{H}'\boldsymbol{\theta} - \boldsymbol{\theta}_0)'\,\mathbf{D}^{-1}(\mathbf{H}'\boldsymbol{\theta} - \boldsymbol{\theta}_0)$$
$$= \sigma^2, \tag{1.3}$$

and since $(\mathbf{H}'\boldsymbol{\theta} - \boldsymbol{\theta}_0)'\,\mathbf{D}^{-1}(\mathbf{H}'\boldsymbol{\theta} - \boldsymbol{\theta}_0)$ is positive definite, (1.3) exceeds σ^2 if and only if the null hypothesis is not true. On the other hand, the expected value of $s_e^2 = [R_0^2/(n-m)]$ is σ^2 without any assumption on the unknown parameter $\boldsymbol{\theta}$. Consider now the statistic defined by

$$F = (\mathbf{Z} - \boldsymbol{\theta}_0)'\,\mathbf{D}^{-1}(\mathbf{Z} - \boldsymbol{\theta}_0)/r \div s_e^2. \tag{1.4}$$

The numerator in (1.4) is, on the average, larger in magnitude than the denominator if the null hypothesis is not true, depending on the non null quantity $(\mathbf{H}'\boldsymbol{\theta} - \boldsymbol{\theta}_0)'\;\mathbf{D}^{-1}(\mathbf{H}'\boldsymbol{\theta} - \boldsymbol{\theta}_0)/r$. Hence large values of F indicate departure from null hypothesis. Choosing a value F_α such that

$P(F > F_\alpha) = \alpha$, where F follows a Snedecor's F distribution on r and $n - m$ degrees of freedom. We may set up the test criterion

$$F = \frac{(\mathbf{Z} - \boldsymbol{\theta}_0)'\,\mathbf{D}^{-1}(\mathbf{Z} - \boldsymbol{\theta}_0)}{rs_e^2} > F_\alpha$$

for rejecting the null hypothesis at α level of significance.

An alternative way of obtaining the numerator in F without obtaining the least squares estimates is the following. Define

$$R_1^2 = \min_{\mathbf{H}'\boldsymbol{\theta} = \boldsymbol{\theta}_0} (\mathbf{y} - \mathbf{X}\boldsymbol{\theta})'\,(\mathbf{y} - \mathbf{X}\boldsymbol{\theta}).$$

Then

$$(\mathbf{Z} - \boldsymbol{\theta}_0)'\,\mathbf{D}^{-1}(\mathbf{Z} - \boldsymbol{\theta}_0) = R_1^2 - R_0^2$$

and we have

$$F = \frac{R_1^2 - R_0^2}{r} \div \frac{R_0^2}{n-m}.$$

We may now set up the analysis of variance table.

Table 1.1 ANOVA

Source	D.F.	S.S	M.S.
Deviation from Hypothesis $\mathbf{H}'\boldsymbol{\theta} = \boldsymbol{\theta}_0$	r	$R_1^2 - R_0^2$	$(R_1^2 - R_0^2)/r$
Residual	$n-m$	R_0^2	$R_0^2/(n-m)$
Total	$n-m-r$	R_1^2	

In a block design situation, the interest of the experimenter is seldom in the entire vector of unknown parameters $\boldsymbol{\theta}$. The interest is generally in a subset of parameter vector $\boldsymbol{\theta}$, say $\boldsymbol{\theta}_{(1)}$. In such a situation, the normal equations $\mathbf{X}'\mathbf{X}\boldsymbol{\theta} = \mathbf{X}'\mathbf{y}$ are not of much interest and one has to obtain normal equations pertaining to the parameters of interest.

Consider once again the linear model (1.1). Let $\boldsymbol{\theta} = (\boldsymbol{\theta}_{(1)}' \; \boldsymbol{\theta}'_{(2)})'$, where $\boldsymbol{\theta}_{(1)}$ is a ν-component vector containing all parameters of interest to the experimenter and $\boldsymbol{\theta}_{(2)}$ is a $(k-\nu)$-component vector containing the set of parameters in the model which are not of much interest to the experimenter. Let the matrix $\mathbf{X}$ be partitioned as $[\mathbf{X}_1 : \mathbf{X}_2]$, where the columns of $\mathbf{X}_1$ correspond to the parameters of interest $\boldsymbol{\theta}_{(1)}$ and the columns of $\mathbf{X}_2$ correspond to the parameters not of interest $\boldsymbol{\theta}_{(2)}$. The normal equations $\mathbf{X}'\mathbf{X}\boldsymbol{\theta}=\mathbf{X}'\mathbf{y}$ reduce to

$$\begin{bmatrix} \mathbf{X}_1'\mathbf{X}_1 & \mathbf{X}'_1\mathbf{X}_2 \\ \mathbf{X}_2'\mathbf{X}_1 & \mathbf{X}_2'\mathbf{X}_2 \end{bmatrix} \begin{bmatrix} \boldsymbol{\theta}_{(1)} \\ \boldsymbol{\theta}_{(2)} \end{bmatrix} = \begin{bmatrix} \mathbf{X}_1'\mathbf{y} \\ \mathbf{X}_2'\mathbf{y} \end{bmatrix}.$$

From these equations we obtain, on eliminating $\boldsymbol{\theta}_{(2)}$, the equations involving only $\boldsymbol{\theta}_{(1)}$ as

$$[\mathbf{A}(\boldsymbol{\theta}_{(1)})]\,\boldsymbol{\theta}_{(1)} = \mathbf{Q}^{(1)}_{(y)}, \tag{1.5}$$

where

$$\mathbf{A}\,(\boldsymbol{\theta}_{(1)}) = \mathbf{X}_1'\,\mathbf{X}_1 - \mathbf{X}_1'\,\mathbf{X}_2\,(\mathbf{X}_2'\,\mathbf{X}_2)^-\,\mathbf{X}_2'\,\mathbf{X}_1,$$

$$\mathbf{Q}^{(1)}_{(y)} = \mathbf{X}_1'\,\mathbf{y} - \mathbf{X}_1'\,\mathbf{X}_2\,(\mathbf{X}_2'\,\mathbf{X}_2)^-\,\mathbf{X}_2'\,\mathbf{y}.$$

Equations (1.5) are called the reduced normal equations for $\boldsymbol{\theta}_{(1)}$ obtained by eliminating the effect of $\boldsymbol{\theta}_{(2)}$. $\mathbf{A}\,(\boldsymbol{\theta}_{(1)})$ is the information matrix associated with $\boldsymbol{\theta}_{(1)}$.

2. Analysis of Block Designs

2.1 INTRODUCTION

Block designs are useful in experiments requiring elimination of heterogeneity in one direction. The most commonly used block design is the randomized complete block design wherein the entire experimental material is divided into homogeneous groups, called blocks, and the treatments are allotted randomly to experimental units within blocks. In this design every treatment occurs exactly once in every block. This design has many advantages, particularly in the simplicity of analysis of the data obtained, but this design cannot be adopted in all situations. If the experiment involves large number of treatments so that it becomes difficult to have large blocks and preserve homogeneous conditions within each block, or if the size of the blocks is determined by the nature of the experimental material or technique, then incomplete block designs are used. An incomplete block design is one in which all the treatments do not occur in at least one block. Incomplete block designs may be binary or p-ary, with varying replications and varying block sizes and may even have more number of plots in a block than the number of treatments in the design. A consequence of using such designs is that the analysis of the data generated through these designs becomes complicated. Therefore, many efforts made in developing this subject were focussed at obtaining designs with patterns which simplify the analysis of the data generated through such designs and also enable important contrasts to be estimated and tested with high efficiency. Characterization and construction of designs having patterns admitting simplified analysis and possessing desirable statistical properties thus became a priority of research workers. In this Chapter we present a general procedure of intra-block analysis of block designs and study some important characterizations.

2.2 FRAMEWORK OF BLOCK DESIGNS

Let us first describe a general procedure of intra-block analysis of block designs. A block design may be represented as $D(v, b, \mathbf{r}, \mathbf{k})$, where v treatments are arranged in b blocks such that $\mathbf{r} = (r_1, \ldots, r_v)'$, is a v-component vector of treatment replications and $\mathbf{k} = (k_1, \ldots, k_b)'$, is a b-component vector of block sizes. The $v \times b$ treatment – block incidence matrix $\mathbf{N}$ has elements n_{ij} $(i = 1, \ldots, v;\ j = 1, \ldots, b)$, where n_{ij} can assume any of the non-negative integral values $x_1, \ldots, x_p$ (say), not all x_i ($i = 1, \ldots, p$) being zero simultaneously, such that $x_i = 0, 1, \ldots, p - 1$, for all i, and n is the total number of observations obtained from the design. Designs with $\mathbf{r} = r\,\mathbf{1}$ are equireplicate and with $\mathbf{k} = k\,\mathbf{1}$ are proper. If n_{ij} has two values, say 0 or 1, the design is binary; if it assumes three values 0, 1, 2, then it is ternary and so on; in general when n_{ij} assumes values 0, 1, ..., p-1, then the design is called p-ary.

A proper and equi-replicate block design $D(v, b, r, k)$ is said to be α-resolvable if the b blocks can be divided into m groups containing n blocks each such that in each group every treatment appears α times. Clearly $v\alpha = k\,n$, $b = mn$, $r = m\alpha$. The above definition is due to Shrikhande and Raghavarao (1963). When $\alpha = 1$, we get 1-resolvable or simply resolvable design of Bose (1942). The concept of resolvability was further extended by Bose to affine resolvability and by Shrikhande and Raghavarao to affine α-resolvability. An α-resolvable design is called affine α-resolvable if any pair of blocks of the same group intersects in q_1 treatments and any pair of blocks from different groups intersects in q_2 treatments. Further, in an affine α-resolvable design k^2/v is an integer.

2 3 ANALYSIS OF BLOCK DESIGNS

The intra-block linear model of n observations obtained from the design $D\,(v, b, \mathbf{r}, \mathbf{k})$ can be written as

$$\begin{aligned} y_{ijk} &= \mu + \tau_i + \beta_j + e_{ijk}, \\ (i &= 1, \ldots, v;\ j = 1, \ldots, b; \\ k &= 1, \ldots, n_{ij}), \end{aligned} \tag{2.1}$$

where y_{ijk} is the k-th observation from i-th treatment in j-th block, μ is the general mean, τ_i is the i-th treatment effect, β_j is the j-th block effect and e_{ijk}'s are residuals distributed identically and independently satisfying

$$E\,(e_{ijk}) = 0,\ V\,(e_{ijk}) = \sigma^2$$

and

$$\text{Cov}\,(e_{ijk}, e_{i'j'k'}) = 0, \text{ for all } ijk \neq i'j'k'.$$

The model (2.1) can also be expressed as

$$\mathbf{y} = \mu\,\mathbf{1} + \boldsymbol{\Delta}'\,\boldsymbol{\tau} + \mathbf{D}'\,\boldsymbol{\beta} + \mathbf{e}, \tag{2.2}$$

with n-component residual vector $\mathbf{e} = (e_1, \ldots, e_n)'$ satisfying $E\,(\mathbf{e}) = \mathbf{0}$ and $D\,(\mathbf{e}) = \sigma^2 I_n$.

The model (2.2) can equivalently be rewritten as

$$\mathbf{y} = \mathbf{X}\,\boldsymbol{\theta} + \mathbf{e}. \tag{2.3}$$

Here $\mathbf{y} = (y_1, \ldots, y_n)'$ is an n component vector of observations, $\mathbf{X} = [\mathbf{1} \,\vdots\, \boldsymbol{\Delta}' \,\vdots\, \mathbf{D}']$ is the $n\times(1 + v + b)$ design matrix partitioned as n-component vector $\mathbf{1}$ of all elements unity, $\boldsymbol{\Delta}'$ an $n\times v$ design matrix for treatments and

$\mathbf{D}'$ an $n \times b$ design matrix for blocks and $\boldsymbol{\theta} = [\mu : \boldsymbol{\tau}' : \boldsymbol{\beta}']'$, with $\boldsymbol{\tau} = (\tau_1, \ldots, \tau_v)'$ and $\boldsymbol{\beta} = (\beta_1, \ldots, \beta_b)'$. The elements of $\boldsymbol{\Delta}$ and $\mathbf{D}$ matrices are 0 or 1. Also

$$\boldsymbol{\Delta}'\mathbf{1} = \mathbf{1},\ \mathbf{D}'\,\mathbf{1} = \mathbf{1},\ \boldsymbol{\Delta}\,\mathbf{1} = \mathbf{r},\ \mathbf{D}\,\mathbf{1} = \mathbf{k},$$

$$\boldsymbol{\Delta}'\boldsymbol{\Delta} = \mathbf{I},\ \mathbf{D}'\mathbf{D} = \mathbf{I}.$$

Moreover, $\mathbf{y} \sim N_n(\mathbf{X}\,\boldsymbol{\theta}, \sigma^2\,\mathbf{I}_n)$.

The following important relations are easily established:

$$\mathbf{N} = \boldsymbol{\Delta}\,\mathbf{D}', \qquad \mathbf{N}\,\mathbf{1} = \mathbf{r}, \qquad \mathbf{N}'\mathbf{1} = \mathbf{k}, \qquad \mathbf{r}'\mathbf{1} = n = \mathbf{k}'\,\mathbf{1},$$

$\mathbf{R} = \text{diag}\,(r_1, \ldots, r_v) = \boldsymbol{\Delta}\,\boldsymbol{\Delta}'$, and $\mathbf{K} = \text{diag}\,(k_1, \ldots, k_b) = \mathbf{D}\,\mathbf{D}'$. We also have $\mathbf{T} = \boldsymbol{\Delta}\,\mathbf{y}$, a vector of treatment totals, $\mathbf{B} = \mathbf{D}\,\mathbf{y}$, a vector of block totals, and $\mathbf{1}'\mathbf{T} = \mathbf{1}'\mathbf{B} = G = \mathbf{1}'\mathbf{y}$, the grand total.

2.3.1 Reduced normal equations

The set of $(1 + v + b)$ normal equations $\mathbf{X}'\,\mathbf{X}\,\boldsymbol{\theta} = \mathbf{X}'\mathbf{y}$ are

$$\begin{bmatrix} n & \mathbf{r}' & \mathbf{k}' \\ \mathbf{r} & \mathbf{R} & \mathbf{N} \\ \mathbf{k} & \mathbf{N}' & \mathbf{K} \end{bmatrix} \begin{bmatrix} \mu \\ \boldsymbol{\tau} \\ \boldsymbol{\beta} \end{bmatrix} = \begin{bmatrix} G \\ \mathbf{T} \\ \mathbf{B} \end{bmatrix}. \tag{2.4}$$

Rewriting the model (2.2) as

$$\mathbf{y} = \boldsymbol{\Delta}'\boldsymbol{\tau} + \mu\mathbf{1} + \mathbf{D}'\boldsymbol{\beta} + \mathbf{e},$$

$$E\,(\mathbf{e}) = \mathbf{0} \text{ and } D\,(\mathbf{e}) = \sigma^2\,\mathbf{I}_n,$$

so as to bring the parameters of interest first and then the parameters not of interest, we can rewrite the normal equations as

$$\begin{bmatrix} \mathbf{R} & \mathbf{r} & \mathbf{N} \\ \mathbf{r}' & n & \mathbf{k}' \\ \mathbf{N}' & \mathbf{k} & \mathbf{K} \end{bmatrix} \begin{bmatrix} \boldsymbol{\tau} \\ \mu \\ \boldsymbol{\beta} \end{bmatrix} = \begin{bmatrix} \mathbf{T} \\ G \\ \mathbf{B} \end{bmatrix}. \tag{2.5}$$

Using (1.5), the reduced normal equations for studying the treatment effects are

$$\mathbf{A}\,\boldsymbol{\tau} = \mathbf{Q}, \tag{2.6}$$

where $\mathbf{A} = \mathbf{R} - \mathbf{N}\mathbf{K}^{-1}\,\mathbf{N}'$, is the information matrix and $\mathbf{Q} = \mathbf{T} - \mathbf{N}\mathbf{K}^{-1}\,\mathbf{B}$, is the vector of adjusted treatment totals.

It is easy to prove the following theorem.

Theorem 2.1: i) $E\,(\mathbf{Q}) = \mathbf{A}\boldsymbol{\tau}$

ii) $D\,(\mathbf{Q}) = \sigma^2\mathbf{A}$.

Proof: The proof follows by writing

$$\mathbf{Q} = (\boldsymbol{\Delta} - \mathbf{N}\mathbf{K}^{-1}\mathbf{D})\,\mathbf{y}.$$

The reduced normal equations (2.6) and the information matrix **A** find prominent place in the analysis of block designs. The readers may, therefore, familiarize themselves with the notations used so far.

2.3.2 Properties of reduced normal equations and A–matrix

The following results are important:

Lemma 2.1: The matrix **A** is singular.

Proof: The singularity is established by noting the fact that $\mathbf{A1} = \mathbf{0}$.

Remark 2.1: Since $\mathbf{A1} = \mathbf{0}$, **A** has **1** as an eigen vector corresponding to its zero eigen value.

Lemma 2.2: $\mathbf{1}'\,\mathbf{Q} = 0 = \mathbf{Q}'\,\mathbf{1}$.

Proof is left as an exercise.

Lemma 2.3: The system of reduced normal equations is consistent.

Proof: We have $\mathbf{1}'\,\mathbf{Q} = 0$ and $\mathbf{1}'\,\mathbf{A} = \mathbf{0}'$. It follows from here that **Q** belongs to the linear manifold of columns of **A**, i.e. $\mathbf{Q} \in \eta\,(\mathbf{A})$. Hence $R(\mathbf{A}) = R\,(\mathbf{A} \vdots \mathbf{Q})$. Hence the system of reduced normal equations is consistent.

Remark 2.2: The system of reduced normal equations is consistent and there always exists a solution. However, there does not exist a unique solution $\hat{\boldsymbol{\tau}}$ of $\boldsymbol{\tau}$, because the matrix **A** is singular. A linear parametric function $\mathbf{p}'\,\boldsymbol{\tau}$ is estimable, however, if $\mathbf{p} \in \eta\,(\mathbf{A})$.

Definition 2.1: A parametric function $\mathbf{p}'\boldsymbol{\tau}$ is a contrast if $\mathbf{p}'\mathbf{1} = 0$. If **p** is such that there are only two non-zero elements 1 and -1 of **p**, then $\mathbf{p}'\boldsymbol{\tau}$ is called an elementary contrast.

Definition 2.2: Two contrasts $\mathbf{p}'\,\boldsymbol{\tau}$ and $\mathbf{q}'\,\boldsymbol{\tau}$ are orthogonal if $\mathbf{p}'\,\mathbf{q} = 0$. If there are more than two contrasts then they are said to form a set of mutually orthogonal contrasts if every pair of the set is orthogonal. If further $\mathbf{p}'\,\mathbf{p} = 1$ then $\mathbf{p}'\,\boldsymbol{\tau}$ is said to be a normalized contrast. A set of mutually orthogonal normalized contrasts is called a set of orthonormal contrasts.

The matrix **A** is singular and its rank can atmost be $(v-1)$. Since there are v treatment parameters, there are $(v-1)$ linearly independent contrasts. In design of experiments, the interest of the experimenter lies in making various comparisons among treatment effects. It is, therefore, desirable that all the $(v-1)$ linearly independent contrasts (and hence all the elementary contrasts) are estimable through the design. This is possible if and only if $R(\mathbf{A})=(v-1)$. This can be ensured by using a design property called connectedness.

Definition 2.3: A design is said to be connected if all the elementary contrasts are estimable through the design. An equivalent definition is as follows. A design is connected if, for any two given treatments α and θ, there exists a chain of treatments $\alpha = \alpha_1, \alpha_2, \ldots, \alpha_v = \theta$ such that every consecutive pair of treatments in the chain appears together in a block. According to Bose (1947a), a block and treatment are said to be associated if the treatment occurs in the block. Two treatments are then said to be connected if a chain of treatments and blocks can be formed as, treatment—block—treatment—block—...—treatment, beginning with α and ending with θ such that every block is associated with both treatments adjacent to it. A design is connected if every pair of treatments is connected.

We now prove the following theorem:

Theorem 2.2: A block design is connected if and only if the rank of its $\mathbf{A}$ matrix is $(v-1)$.

Proof: Let the design be connected and let $\mathbf{H}'\boldsymbol{\tau}$ be a set of linearly independent estimable contrasts, where $\mathbf{H}$ is a $v \times (v-1)$ matrix. Here $R(\mathbf{H})=(v-1)$. Since $\mathbf{H}'\boldsymbol{\tau}$ is estimable, $\mathbf{H} \in \eta(\mathbf{A})$ or $\mathbf{H} = \mathbf{A}\,\mathbf{C}$, where $\mathbf{C}$ is a $v\times(v-1)$ matrix. Now $v-1 = R(\mathbf{H}) \leqslant R(\mathbf{A})$. But $R(\mathbf{A}) \leqslant v-1$. Hence $R(\mathbf{A}) = v-1$.

Conversely, let us assume that the matrix $\mathbf{A}$ is of rank $(v-1)$. Let $\boldsymbol{\xi}_1, \ldots, \boldsymbol{\xi}_{v-1}$ be a set of orthonormal eigen vectors corresponding to the non-zero eigen values $\theta_1, \ldots, \theta_{v-1}$ of $\mathbf{A}$.
Then

$$\mathbf{A} = \sum_{i=1}^{v-1} \theta_i \boldsymbol{\xi}_i \boldsymbol{\xi}_i' \text{ and } \mathbf{A}^- = \sum_{i=1}^{v-1} \theta_i^{-1} \boldsymbol{\xi}_i \boldsymbol{\xi}_i'.$$

$$E(\boldsymbol{\xi}_i'\hat{\boldsymbol{\tau}}) = E[\boldsymbol{\xi}_i'(\sum_{i=1}^{v-1} \theta_i^{-1}\boldsymbol{\xi}_i\boldsymbol{\xi}_i')\mathbf{Q}] = \boldsymbol{\xi}_i'\boldsymbol{\tau}.$$

Thus $\boldsymbol{\xi}_i'\boldsymbol{\tau}$ $(i = 1, \ldots, v-1)$ are estimable and are estimated by

$$\boldsymbol{\xi}_i'\hat{\boldsymbol{\tau}} = \boldsymbol{\xi}_i'\mathbf{Q}/\theta_i.$$

For some arbitrary choice of $a_1, \ldots, a_{v-1}$, let

$$\mathbf{p} = \sum_{i=1}^{v-1} a_i \boldsymbol{\xi}_i.$$

Then

$$E(\mathbf{p}'\hat{\boldsymbol{\tau}}) = E[\sum_{i=1}^{v-1} a_i \boldsymbol{\xi}_i' \hat{\boldsymbol{\tau}}] = \mathbf{p}'\boldsymbol{\tau}$$

Thus $\mathbf{p}'\boldsymbol{\tau}$ is estimable.

Alternatively, the converse may be proved by noting that if

$$R(\mathbf{A}) = v - 1,$$

then for every $v \times 1$ vector $\mathbf{p}$ with $\mathbf{p}'\mathbf{1} = 0$, one has $\mathbf{p} \in$ row space $(\mathbf{A})$.

The proof is thus complete.

Throughout we shall assume that the design is connected so that

$$R(\mathbf{A}) = v - 1.$$

2.3.3 Analysis of variance

It is seen that although the matrix $\mathbf{A}$ is singular the system of reduced normal equations (2.6) is consistent. A particular solution of $\boldsymbol{\tau}$ is then

$$\hat{\boldsymbol{\tau}} = \mathbf{A}^{-} \mathbf{Q}.$$

From (2.5) we have

$$\hat{\boldsymbol{\beta}} = \mathbf{K}^{-1} \mathbf{B} - \hat{\mu}\mathbf{1} - \mathbf{K}^{-1} \mathbf{N}'\hat{\boldsymbol{\tau}}.$$

Let one particular set of solutions be

$$\hat{\boldsymbol{\theta}}' = [\hat{\mu} \vdots \hat{\boldsymbol{\tau}}' \vdots (\mathbf{K}^{-1}\mathbf{B} - \hat{\mu}\mathbf{1} - \mathbf{K}^{-1}\mathbf{N}'\hat{\boldsymbol{\tau}})'].$$

Then

$$R_0^2 = \mathbf{y}'\mathbf{y} - \hat{\boldsymbol{\theta}}' \mathbf{X}' \mathbf{Y}$$
$$= \mathbf{y}'\mathbf{y} - \mathbf{Q}'\hat{\boldsymbol{\tau}} - \mathbf{B}'\mathbf{K}^{-1}\mathbf{B}.$$

In fact $R_0^2/(n - v - b + 1)$ is an unbiased estimator of σ^2.

We are interested in testing the null hypothesis

$$H_0: \quad \tau_1 = \tau_2 = \ldots = \tau_v = \tau, \text{ (say)}$$

or

$$H_0: \quad \boldsymbol{\tau} = \tau\mathbf{1},$$

or

$$H_0: \quad \tau_1 - \tau_i = 0, \; i = 2, \ldots, v,$$

against the alternative that not all τ_i's $(i = 1, \ldots, v)$ are equal.

Under the null hypothesis, the model (2.2) reduces to

$$\mathbf{y} = \alpha\mathbf{1} + \mathbf{D}'\boldsymbol{\beta} + \mathbf{e}, \tag{2.7}$$
$$E(\mathbf{e}) = \mathbf{0} \text{ and } D(\mathbf{e}) = \sigma^2 \mathbf{I}_n$$

where $\alpha = \mu + \tau$.

For the model (2.7) the residual sum of squares is given by

$$R_1^2 = \mathbf{y}'\mathbf{y} - \mathbf{B}'\mathbf{K}^{-1} \mathbf{B}$$

and the block sum of squares under (2.7) is $\mathbf{B}'\mathbf{K}^{-1} \mathbf{B} - G^2/n$.

We thus have

$$R_1^2 - R_0^2 = \mathbf{Q}'\hat{\boldsymbol{\tau}}.$$

The term $\mathbf{Q}'\hat{\boldsymbol{\tau}} = \mathbf{Q}'\mathbf{A}^-\mathbf{Q}$ represents the reduction in the sum of squares of residuals due to adding the treatment effects to the model and is called the treatment sum of squares adjusted for blocks. The null hypothesis H_0 is tested using the statistic

$$F = [\mathbf{Q}'\mathbf{A}^-\mathbf{Q}/(v-1)] \div [R_0^2/(n-v-b+1)]$$

which has the Snedecor's F distribution with $(v-1)$ and $(n-v-b+1)$ degrees of freedom.

Let $\mathbf{H}'\boldsymbol{\tau}$ be a set of $(v-1)$ linearly independent parametric functions. The dispersion matrix of $\mathbf{H}'\hat{\boldsymbol{\tau}}$, the best linear unbiased estimator of $\mathbf{H}'\boldsymbol{\tau}$, is

$$D(\mathbf{H}'\hat{\boldsymbol{\tau}}) = D(\mathbf{H}'\mathbf{A}^-\mathbf{Q})$$
$$= \sigma^2 \mathbf{H}'\mathbf{A}^-\mathbf{H}.$$

In particular, let $\mathbf{A}^- = (a^{ij})$,. Then

$$V(\widehat{\tau_i - \tau_j}) = \sigma^2 [a^{ii} - 2a^{ij} + a^{jj}).$$

2.4 THE $\boldsymbol{\Omega}^{-1}$ MATRIX

It has been shown earlier that the information matrix $\mathbf{A}$ in the reduced normal equations (2.6) is singular and has rank $(v-1)$ if the design is connected. In order to obtain the estimates of estimable parametric functions (or the elementary treatment contrasts), a generalized inverse $\mathbf{A}^-$ of $\mathbf{A}$ is required. A generalized inverse is not unique and can be obtained in a number of ways. Tocher (1952) considered an important matrix, $\boldsymbol{\Omega}^{-1}$, defined as

$$\boldsymbol{\Omega}^{-1} = \mathbf{A} + \mathbf{r}\,\mathbf{r}'/n \qquad (2.8)$$
$$= \mathbf{R} - \mathbf{N}\,\mathbf{K}^{-1}\,\mathbf{N}' + \mathbf{r}\,\mathbf{r}'/n.$$

For connected designs the matrix $\boldsymbol{\Omega}^{-1}$ is non-singular and $\boldsymbol{\Omega}$ is a g-inverse of $\mathbf{A}$, and a solution of the reduced normal equations is $\hat{\boldsymbol{\tau}} = \boldsymbol{\Omega}\,\mathbf{Q}$, where $\boldsymbol{\Omega}$ is the inverse of $\boldsymbol{\Omega}^{-1}$. The treatment sum of squares adjusted for block effects is given by $\mathbf{Q}'\,\boldsymbol{\Omega}\,\mathbf{Q}$. An adjusted treatment mean is obtained by adding the general mean $m = G/n$ to the appropriate element of $\hat{\boldsymbol{\tau}}$, i.e. $\mathbf{a} = m\,\mathbf{1} + \boldsymbol{\Omega}\,\mathbf{Q}$, where $\mathbf{a}$ denotes the vector of adjusted treatment means.

Further, $\sigma^2\,\boldsymbol{\Omega}$ is the pseudo dispersion matrix of the estimated treatment parameters $\hat{\boldsymbol{\tau}}$, so that if $\mathbf{p}'\,\boldsymbol{\tau}$ is an estimable parametric function, then $\mathbf{p}'\,\hat{\boldsymbol{\tau}}$ has a standard error of $[(\mathbf{p}'\,\boldsymbol{\Omega}\,\mathbf{p})\,\sigma^2]^{1/2}$.

Remark 2.3: It is easy to verify that $\boldsymbol{\Omega}^{-1}\,\mathbf{1} = \mathbf{r}$, so that $\boldsymbol{\Omega}\,\mathbf{r} = \mathbf{1}$.

It may be remarked here that the $\boldsymbol{\Omega}^{-1}$ matrix of Tocher is not the only way to obtain $\mathbf{A}^-$. John (1965) used corollary 1.2 of Theorem 1.1 to obtain a generalized inverse of $\mathbf{A}$.

By suitable choices of q and $\mathbf{Z}$, $\mathbf{\Omega}^{-1}$ can usually be expressed in a form which allows for easy inversion. John and Smith (1972) preferred choosing the constraint $\mathbf{Z} = \mathbf{1}$. Tocher (1952) defined $\mathbf{\Omega}^{-1}$ in the special case of $\mathbf{Z} = \mathbf{r}$ and $q = 1/n$. We shall restrict ourselves to this $\mathbf{\Omega}^{-1}$ only.

The problem now is to obtain $\mathbf{\Omega}$, the inverse of $\mathbf{\Omega}^{-1}$. If $\mathbf{\Omega}^{-1}$ is highly structured then $\mathbf{\Omega}$ is also structured and can be determined easily. However, the main concern has been when $\mathbf{\Omega}^{-1}$ matrix does not have a structure. Some iterative procedures of analysis are available in the literature for this situation and a review of these is in order.

2.5 INVERSION OF $\mathbf{\Omega}^{-1}$

The iterative method for the analysis of non-orthogonal designs was used by Stevens (1948) and was later formalized by Corsten (1958). Wilkinson (1970) proposed a method of wide application for the analysis of experimental data. The method essentially involves sweeps to eliminate the effects of parameters and eventually to leave an estimate of $\boldsymbol{\tau}$. Calinski (1971) advanced an iterative formula for obtaining $\mathbf{\Omega}$, the inverse of $\mathbf{\Omega}^{-1}$ matrix, introduced by Tocher (1952), for obtaining estimates of $\boldsymbol{\tau}$.

In deriving a general formula for $\mathbf{\Omega}$, the following matrices will be utilized:

$$\mathbf{N}_0 = \mathbf{N} - \mathbf{r}\,\mathbf{k}'/n,$$

used by Rees (1966) and Pearce (1970);

$$\mathbf{P} = \mathbf{N}\,\mathbf{K}^{-1}\,\mathbf{N}';$$

$$\mathbf{P}_0 = \mathbf{N}_0\,\mathbf{K}^{-1}\,\mathbf{N}_0' = \mathbf{P} - \mathbf{r}\,\mathbf{r}'/n;\ \mathbf{M} = \mathbf{R}^{-1}\,\mathbf{P} = \mathbf{R}^{-1}\,\mathbf{N}\mathbf{K}^{-1}\,\mathbf{N}',$$

examined by Jones (1959),

$$\mathbf{M}_0 = \mathbf{R}^{-1}\,\mathbf{P}_0 = \mathbf{R}^{-1}\,\mathbf{P} - \mathbf{1}\,\mathbf{r}'/n = \mathbf{R}^{-1}\,\mathbf{N}\,\mathbf{K}^{-1}\,\mathbf{N}' - \mathbf{1}\,\mathbf{r}'/n.$$

Notice also that

$$\mathbf{P}\,\mathbf{1} = \mathbf{r},\ \mathbf{M}\,\mathbf{1} = \mathbf{1},\ \mathbf{P}_0\,\mathbf{1} = \mathbf{0}$$

and hence

$$\mathbf{M}_0\,\mathbf{1} = \mathbf{0}.$$

The method of obtaining $\mathbf{\Omega}$ matrix given here is due to Dey and Gupta (1986). We first give some definitions and state some Lemmas:

Definition 2.4: Two square matrices $\mathbf{A}$ and $\mathbf{B}$ of the same order are said to be similar if $\mathbf{A} = \mathbf{S}^{-1}\,\mathbf{B}\,\mathbf{S}$ for some non-singular matrix $\mathbf{S}$.

Lemma 2.4: If **A** and **B** are similar matrices, then $R(\mathbf{A}) = R(\mathbf{B})$. Further, the eigen values of **A** and **B** are same including multiplicities.

Definition 2.5: A square matrix **A** with non-negative elements is said to be stochastic if $\mathbf{A1} = \mathbf{1}$.

Lemma 2.5: For a stochastic matrix **A**, one of the eigen values is unity and the other eigen values have modulii less than or equal to unity.

For a proof, see e.g. Bartlett (1978, p. 28).

Lemma 2.6: If the modulus of each eigen value of a square matrix **A** is less than unity, then the series $\mathbf{I} + \mathbf{A} + \mathbf{A}^2 + \ldots$ converges to $(\mathbf{I} - \mathbf{A})^{-1}$.

For a proof, see Mirsky (1972, p. 332).

We now obtain the $\boldsymbol{\Omega}$ matrix. We have

$$\boldsymbol{\Omega}^{-1} = \mathbf{R} - \mathbf{NK}^{-1}\mathbf{N}' + \mathbf{rr}'/n = \mathbf{R}\ (\mathbf{I} - \mathbf{M}_0). \tag{2.9}$$

We also have

$$\mathbf{M} = \mathbf{R}^{-1}\mathbf{NK}^{-1}\mathbf{N}' = \mathbf{R}^{-1/2}\mathbf{R}^{-1/2}\mathbf{NK}^{-1}\mathbf{N}'\mathbf{R}^{-1/2}\mathbf{R}^{1/2}$$
$$= \mathbf{R}^{-1/2}\mathbf{BR}^{1/2}, \tag{2.10}$$

where $\mathbf{B} = \mathbf{R}^{-1/2}\mathbf{NK}^{-1}\mathbf{N}'\mathbf{R}^{-1/2} = \mathbf{B}'$.

Also notice that $\mathbf{B}\,\mathbf{r}^{1/2} = \mathbf{r}^{1/2}$. (2.11)

We then have the following:

Lemma 2.7: The eigen values of **M** and **B** are same, including multiplicities, and are therefore all real.

Proof: Follows from Lemma 2.4.

Lemma 2.8: The eigen values of **M** are $\mu_0 = 1, \mu_1, \ldots, \mu_s$ with multiplicities $\rho_0 = 1, \rho_1, \ldots, \rho_s$ respectively, $\sum_{i=1}^{s} \rho_i = v - 1$, where $0 \leqslant \mu_i < 1$, $i = 1, \ldots, s$.

Proof: Since the elements of the matrix **M** are all non-negative and $\mathbf{M1} = \mathbf{1}$, it follows from definition 2.5 that **M** is a stochastic matrix. Using Lemma 2.5, one of the eigen values of **M** is unity and other eigen values of **M** have modulii less than or equal to unity, i.e. $|\mu_i| \leqslant 1$. From Lemma 2.7, since **B** is non-negative definite, it follows that $0 \leqslant \mu_i \leqslant 1$, for all $i = 1, \ldots, s$. Also $\mathbf{M}_0 = \mathbf{M} - \mathbf{1r}'/n$. Since **M** and $\mathbf{1r}'/n$ commute, the eigen values of $\mathbf{M}_0$ are the difference of the eigen values of **M** and $\mathbf{1\,r}'/n$. But $\mathbf{1\,r}'/n$ has only one eigen value as 1. Therefore, the eigen values of $\mathbf{M}_0$ are 0 and μ_i, $i = 1, \ldots, s$. But for a connected design $(\mathbf{I} - \mathbf{M}_0)$ is non-singular. Thus $\mathbf{M}_0$ cannot have any eigen value equal to unity. Hence, $0 \leqslant \mu_i < 1$, and $\mu_0 = 1$ is a simple eigen value of **M** for connected block designs.

Lemma 2.9: For a connected block design,

$$\boldsymbol{\Omega} = (\mathbf{I} + \sum_{h=1}^{\infty} \mathbf{M}_0^h)\, \mathbf{R}^{-1}. \tag{2.12}$$

Proof: From Lemma 2.8, we see that $\mathbf{M}_0$ has eigen values each of which has modulus less than unity. Thus, from Lemma 2.6 and 2.9 the result follows.

The result (2.12) is the same as obtained by Pearce (1970) in the form

$$\boldsymbol{\Omega} = \mathbf{R}^{-1/2}\, [\mathbf{I} + \sum_{h=1}^{\infty} \mathbf{H}^h]\, \mathbf{R}^{-1/2}, \tag{2.13}$$

where $\mathbf{H} = \mathbf{R}^{-1/2}\, \mathbf{P}_0\, \mathbf{R}^{-1/2}$.

The particular behaviour of the $\mathbf{M}_0$ matrix has attracted more attention to the algebraic theory of block designs. Interestingly, although $\mathbf{M}_0$ is asymmetric, it always admits a decomposition (this we prove in Chapter 3)

$$\mathbf{M}_0 = \sum_{i}^{v-1} \mu_i\, \mathbf{L}_i, \tag{2.14}$$

where μ_i's are the distinct eigen values of $\mathbf{M}_0$ and $\mathbf{L}_i$'s are proper idempotent matrices satisfying $\mathbf{L}_i\mathbf{L}_j = \mathbf{0}$, $i \neq j$, and $\sum_{i=1}^{v-1} \mathbf{L}_i = \mathbf{I} - \mathbf{1}\,\mathbf{r}'/n$. This gives

$$\mathbf{M}_0^h = \Sigma_i\, \mu_i^h\, \mathbf{L}_i. \tag{2.15}$$

Using (2.15), the expression in (2.12) simplifies to

$$\boldsymbol{\Omega} = [\mathbf{I} + \Sigma_i\, \mu_i\, \mathbf{L}_i/(1 - \mu_i)]\, \mathbf{R}^{-1}. \tag{2.16}$$

For equireplicate designs, (2.16) further simplifies to

$$\boldsymbol{\Omega} = [\mathbf{I} + \sum_i \mu_i\, \mathbf{L}_i/(1 - \mu_i)]/r.$$

It is quite clear now from the algebra derived above that $\boldsymbol{\Omega}$ matrix can be computed easily once the eigen values μ_i and the corresponding idempotent matrices $\mathbf{L}_i$ of the matrix $\mathbf{M}_0$ are determined. The matrix $\mathbf{M}_0$ thus plays an important role in the analysis and characterization of block designs. The significance of $\mathbf{M}_0$ matrix was emphasized by Calinski (1971). It will be shown in Chapter 3 that the eigen values μ_i are the relative losses of information and hence terms $1 - \mu_i$ determine the efficiency factors of the contrasts and the design. The simplification of the general formula depends entirely on the pattern of this matrix, which, in turn, is helpful in designing block experiments with specific desirable properties. Clearly, for an orthogonal design $\mathbf{M}_0 = \mathbf{0}$, and $\boldsymbol{\Omega} = \mathbf{R}^{-1}$. All non-orthogonal designs can therefore be classified according to the patterns of $\mathbf{M}_0$.

2.6 CLASSIFICATION OF DESIGNS

The first attempt to classify various designs was made by Pearce (1963) utilizing the patterns of the matrix $\mathbf{\Omega}^{-1}$. If the designs are proper, i.e. $\mathbf{k} = k\,\mathbf{1}$, $\mathbf{N}\,\mathbf{K}^{-1}\,\mathbf{N}'$ equals $\mathbf{N}\,\mathbf{N}'/k$. If further the design is equi-replicated, $\mathbf{r}\,\mathbf{r}'/n$ will impart no pattern to $\mathbf{\Omega}^{-1}$, so classifications on the basis of the off-diagonal elements of $\mathbf{N}\,\mathbf{N}'$ will be the same as proposed by Pearce (1963). If, however, the treatments have varying replications in the design, the resulting pattern of $\mathbf{r}\,\mathbf{r}'/n$ will usually reinforce a pattern already in $\mathbf{N}\,\mathbf{N}'/k$. Hence, the characterization of designs has to be made on the basis of $\mathbf{\Omega}^{-1}$ matrix.

We present here some of the useful classifications and characterizations based on $\mathbf{\Omega}^{-1}$ matrix and defer to Chapter 3 some further characterizations based on $\mathbf{M}_0$ and $\mathbf{P} = \mathbf{N}\,\mathbf{K}^{-1}\,\mathbf{N}'$ matrices.

2.6.1. Orthogonal designs

We first prove the following theorem concerning orthogonal designs:

Theorem 2.3: A block design is said to be orthogonal if and only if its incidence matrix $\mathbf{N}$ is of the form

$$\mathbf{N} = \mathbf{r}\mathbf{k}'/n.$$

Proof: By definition, a design is orthogonal if the covariance between the adjusted treatment totals and the adjusted block totals is zero. The vector of adjusted treatment totals is $\mathbf{Q} = \mathbf{T} - \mathbf{N}\,\mathbf{K}^{-1}\,\mathbf{B}$ and the vector of adjusted block totals is $\mathbf{Q}^* = \mathbf{B} - \mathbf{N}'\,\mathbf{R}^{-1}\,\mathbf{T}$.

$$\text{Cov}\,(\mathbf{Q}, \mathbf{Q}^*) = \mathbf{0}$$

$$\Rightarrow \text{Cov}\,[(\mathbf{\Delta} - \mathbf{N}\,\mathbf{K}^{-1}\,\mathbf{D})\,\mathbf{y}, (\mathbf{D} - \mathbf{N}'\mathbf{R}^{-1}\,\mathbf{\Delta})\,\mathbf{y}] = \mathbf{0}$$

$$\Rightarrow \sigma^2\,(\mathbf{\Delta} - \mathbf{N}\,\mathbf{K}^{-1}\,\mathbf{D})\,(\mathbf{D} - \mathbf{N}'\,\mathbf{R}^{-1}\,\mathbf{\Delta})' = \mathbf{0}$$

$$\Rightarrow -\sigma^2\,\mathbf{A}\,\mathbf{R}^{-1}\,\mathbf{N} = \mathbf{0}.$$

Let $\mathbf{R}^{-1}\,\mathbf{N} = (\boldsymbol{\alpha}_1 \vdots \boldsymbol{\alpha}_2 \vdots \ldots \vdots \boldsymbol{\alpha}_b)$.

We have $\mathbf{A}\,\boldsymbol{\alpha}_i = \mathbf{0} \;\forall\; i$.

Also $\mathbf{A}\,\mathbf{1} = \mathbf{0}$.

It therefore follows that

$$\boldsymbol{\alpha}_i = m_i\,\mathbf{1}.$$

Now $\mathbf{r}'\,\mathbf{R}^{-1}\,\mathbf{N} = \mathbf{1}'\,\mathbf{N} = \mathbf{k}' = (\mathbf{r}'\boldsymbol{\alpha}_1\,\mathbf{r}'\boldsymbol{\alpha}_2 \ldots \mathbf{r}'\,\boldsymbol{\alpha}_b)$

$$\Rightarrow \mathbf{r}'\,\boldsymbol{\alpha}_i = k_i.$$

But $\boldsymbol{\alpha}_i = m_i \mathbf{1}$

and $\mathbf{r}' \alpha_i = m_i \mathbf{r}' \mathbf{1} = m_i n$

$$\Rightarrow m_i = k_i/n.$$

Therefore $\mathbf{R}^{-1} \mathbf{N} = (\boldsymbol{\alpha}_1 \ldots \boldsymbol{\alpha}_b)$

$$= (k_1/n \, \mathbf{1} \ldots\ldots k_b/n \, \mathbf{1})$$

$$= \mathbf{1} \, \mathbf{k}'/n.$$

Also $\mathbf{R} \mathbf{R}^{-1} \mathbf{N} = \mathbf{R} \mathbf{1} \mathbf{k}'/n$

$$\Rightarrow \mathbf{N} = \mathbf{r} \, \mathbf{k}'/n$$

Conversely, let $\mathbf{N} = \mathbf{r} \, \mathbf{k}'/n$.

Then $\mathbf{A} = \mathbf{R} - \mathbf{N} \mathbf{K}^{-1} \mathbf{N}'$

$$= \mathbf{R} \, (\mathbf{I} - \mathbf{1} \, \mathbf{r}'/n)$$

and $\mathbf{A}^{-} = \mathbf{R}^{-1}$.

Since $\mathbf{A}^{-}$ is a diagonal matrix, all treatment contrasts are estimated without correlations. Hence the proof.

For orthogonal designs $\mathbf{N} = \mathbf{r} \, \mathbf{k}'/n$ and $\boldsymbol{\Omega}^{-1}$ reduces to

$$\boldsymbol{\Omega}^{-1} = \mathbf{R} - \mathbf{N} \, \mathbf{K}^{-1} \mathbf{N}' + \mathbf{r} \, \mathbf{r}'/n = \mathbf{R}.$$

Thus, if all the off-diagonal elements of $\boldsymbol{\Omega}^{-1}$ equal zero, the design is orthogonal. For orthogonal designs, $\boldsymbol{\Omega} = \mathbf{R}^{-1}$.

2.6.2 Totally balanced designs

Another simple form of the matrix $\boldsymbol{\Omega}^{-1}$ is one in which all its diagonal elements are α and all off-diagonal elements are β. Such designs have been termed as totally balanced by Pearce (1963). From the pattern of $\boldsymbol{\Omega}^{-1}$ it is evident that all treatments must be equi-replicated, since

$$\boldsymbol{\Omega}^{-1} \mathbf{1} = \mathbf{r} = [\alpha + \beta \, (v - 1)] \, \mathbf{1},$$

by virtue of $\boldsymbol{\Omega}^{-1}$ being

$$\boldsymbol{\Omega}^{-1} = (\alpha - \beta) \, \mathbf{I} + \beta \, \mathbf{J}.$$

2.6.3 Partially balanced designs

If (i) all diagonal elements of the matrix $\boldsymbol{\Omega}^{-1}$ are equal and (ii) the elements of any row or column of $\boldsymbol{\Omega}^{-1}$ are obtained by permutation of the elements of any other row or column, the design is partially balanced. The characteristic property of permutability of this class of designs does not help in the inversion of $\boldsymbol{\Omega}^{-1}$. However, in many designs beloning to this class, like group divisible, and cyclic, it may be possible to determine $\boldsymbol{\Omega}$ matrix without much difficulty.

2.6.4 Designs with factorial structure

If (i) $v = mt$, treatments form m groups each of t members and (ii) by taking treatments in a suitable order $\mathbf{\Omega}^{-1}$ can be partitioned as

$$\mathbf{\Omega}^{-1} = \{\mathbf{D}\ \mathbf{D}^* \ldots \mathbf{D}^*\},$$

where { } denotes the cyclic pattern, the sub-matrix $\mathbf{D}$ has all the diagonal elements α and all the off-diagonal elements β, and the diagonal elements of $\mathbf{D}^*$ equal α' and the off-diagonal ones β', the design has factorial balance. This class of designs includes all the latin square type of partially balanced designs with association scheme L_2 (Bose, Clatworthy and Shrikhande, 1954) and simple Lattices (Yates, 1936).

2.6.5 Designs with supplemented balance

If $\mathbf{\Omega}^{-1}$ has a structure as

$$\begin{bmatrix} a_0 & \beta_0\mathbf{1}' \\ \beta_0\mathbf{1} & \mathbf{D} \end{bmatrix},$$

where $\mathbf{D}$ has all diagonal elements equal to α and all off-diagonal elements equal to β, then the design is said to have supplemented balance. The treatment giving rise to the distinctive diagonal element a_0 of $\mathbf{\Omega}^{-1}$ is called the supplemented treatment. These designs were considered by Pearce (1960), though with a more restricted definition. The designs of this type are used mainly for comparing treatments and so it is rarely necessary to invert $\mathbf{\Omega}^{-1}$ in full, though, if this is done, $\mathbf{\Omega}$ has the same pattern as $\mathbf{\Omega}^{-1}$. Since $\mathbf{\Omega}^{-1}\mathbf{1} = \mathbf{r}$, all the non-supplementing treatments must be equally replicated.

2.7 DESIGNS OBTAINABLE FROM A GIVEN BLOCK DESIGN

In this Section we discuss about the designs that can be obtained from a given block design. Let $D(v, b, \mathbf{r}, \mathbf{k})$ be a p-ary block design with incidence matrix $\mathbf{N}$. Then the design D^* obtained from D by interchanging the role of treatments and blocks is called the dual of D. Clearly, D^* has the incidence matrix $\mathbf{N}'$ and has the parameters D^* ($v^* = b$, $b^* = v$, $\mathbf{r}^* = \mathbf{k}$, $\mathbf{k}^* = \mathbf{r}$). From a given design D in v treatments another design D^* can be obtained in $v^* < v$ treatments by merging the v treatments in disjoint groups. Again, a new design D^* can be obtained from D by deleting one or more treatments or blocks or some times all the treatments contained in a block. Sometimes, the design D can be enforced either with new treatments or blocks or with both treatments and blocks. The resulting design D^* is discussed in the literature under different nomenclatures of reinforced, augmented and supplemented block designs. If D is binary with incidences 0 or 1 then a new design D^* is obtained by interchanging the incidences 0 and 1. Such a D^* is called the complementary design.

If $\mathbf{N}_1$ and $\mathbf{N}_2$ are the incidence matrices of two block designs D_1 $(v_1, b_1, \mathbf{r}_1, \mathbf{k}_1)$ and D_2 $(v_2, b_2, \mathbf{r}_2, \mathbf{k}_2)$ respectively, then also by using them together new designs can be obtained either by taking the union of two designs or by considering the Kronecker Product or other product of $\mathbf{N}_1$ and $\mathbf{N}_2$. These methods can easily be generalized to more than two designs.

All the methods discussed above mostly lead to new designs and are important in their own right. In view of their usefulness we study these in detail in the Chapters that follow.

3. Balanced and Partially Balanced Designs

3.1 INTRODUCTION

It is seen in Chapter 2 that $\mathbf{\Omega}^{-1}$ matrix plays an important role in the analysis of block designs. The iterative procedure of obtaining the inverse $\mathbf{\Omega}$, given by Calinski (1971), is quite useful in the analysis if $\mathbf{\Omega}^{-1}$ is unstructured. But when $\mathbf{\Omega}^{-1}$ is structured, the iterative procedure may prove to be a wasteful effort as it may be easier to determine the pattern of $\mathbf{\Omega}$ through direct methods. Pearce (1963) classified designs according to the patterns of $\mathbf{\Omega}^{-1}$ and some important patterns were studied in Chapter 2. This Chapter describes in detail some designs for which $\mathbf{\Omega}^{-1}$ is a structured matrix. Balanced and partially balanced designs form the subject matter of this chapter. Designs with factorial structure, designs with supplemented balance and several other useful designs will be studied in the subsequent Chapters.

3.2 BALANCING IN BLOCK DESIGNS

Balance in incomplete block designs is a desirable statistical property. In literature, the term balance has been used in several senses. We restrict our discussions here to only three types of balances viz., variance balance, efficiency balance and pairwise balance. The first two concepts of balance are based on statistical properties whereas the third one is a combinatorial concept. An excellent review paper on the terminological tangle on different concepts of balance is given by Preece (1982).

3.2.1 Variance balance

Definition 3.1: A connected block design $D(v, b, \mathbf{r}, \mathbf{k})$ is variance balanced if it permits the estimation of all normalized treatment contrasts with the same variance. Equivalently, a design is variance balanced if all the elementary contrasts are estimated with same precision.

The following result, originally due to Rao (1958), gives a necessary and sufficient condition for a block design to be variance balanced.

Theorem 3.1: A connected block design is variance balanced if and only if all the non-zero eigen values of the matrix **A** are equal.

Proof: Let $\theta_1,\ldots,\theta_{v-1}$ be the non-zero eigen values of **A** and $\xi_1,\ldots,\xi_{v-1}$, respectively, the corresponding orthonormal eigen vectors, i.e. $\xi_i'\xi_j = \partial_{ij}$, where ∂_{ij} is the Kronecker delta, and $\xi_i'\mathbf{1} = 0$, for all $i = 1,\ldots, v-1$. Thus

$$\mathbf{A} = \sum_{i=1}^{v-1} \theta_i \xi_i \xi_i' \text{ and } \mathbf{A}^- = \sum_{i=1}^{v-1} \theta_i^{-1} \xi_i \xi_i'.$$

Let $\mathbf{H} = [\xi_1,\ldots,\xi_{v-1}]$. Then $\mathbf{H}'\boldsymbol{\tau}$ denotes a set of $(v-1)$ orthonormal estimable treatment contrasts with best linear unbiased estimator $\mathbf{H}'\hat{\boldsymbol{\tau}}$, where $\hat{\boldsymbol{\tau}}$ is a solution of $\mathbf{A}\boldsymbol{\tau} = \mathbf{Q}$. The dispersion matrix of $\mathbf{H}'\hat{\boldsymbol{\tau}}$ is

$$D(\mathbf{H}'\hat{\boldsymbol{\tau}}) = \sigma^2 \mathbf{H}'\mathbf{A}^-\mathbf{H}$$

$$= \sigma^2 \text{ Diag } (\theta_1^{-1},\ldots,\theta_{v-1}^{-1}). \tag{3.1}$$

The average variance of the estimated $(v-1)$ orthonormal contrasts is

$$\sigma^2 \left(\sum_{i=1}^{v-1} \theta_i^{-1}\right) \Big/ (v-1).$$

Now let the design be variance balanced. Then by definition 3.1 all normalised contrasts are estimated with same variance. In particular, the variance of $\xi_i'\hat{\boldsymbol{\tau}}$, from (3.1), is

$$\text{Var } (\xi_i'\hat{\boldsymbol{\tau}}) = \sigma^2 \theta_i^{-1}, \; i = 1,\ldots, v-1.$$

It follows, therefore, that for a variance balanced design $\theta_1 = \theta_2 = \ldots = \theta_{v-1}$.
Conversely, if $\theta_1 = \ldots = \theta_{v-1} = \theta$ (say), each of the orthonormalized treatment contrasts $\xi_i'\boldsymbol{\tau}$ is estimated with the same variance $\sigma^2\theta^{-1}$, and the design is variance balanced.

Corollary 3.1: A connected block design is variance balanced if and only if all the diagonal elements of the matrix **A** are equal and all its off diagonal elements are equal, i.e., the matrix **A** is of the form $a\mathbf{I} + b\,\mathbf{11}'$ and in view of $\mathbf{A1} = \mathbf{0}$, the matrix **A** is of the form $\theta(\mathbf{I} - \mathbf{11}'/v)$, where θ is the unique non-zero eigen value of **A**.

Remark 3.1: In view of $\mathbf{A1} = \mathbf{0}$, the equality of the off-diagonal elements of **A** implies the equality of the diagonal elements. Noting this, Pearce (1964) observed that it is sufficient to ensure the constancy of off-diagonal elements of the matrix **A** for a design to have variance balance.

Remark 3.2: The result of remark 3.1 was further improved by Puri and Nigam (1976) who characterized variance balance in terms of **P**-matrix. A necessary and sufficient condition for a block design to be variance balanced is that $\mathbf{P} = \mathbf{N}\,\mathbf{K}^{-1}\,\mathbf{N}'$ has all its off diagonal elements equal, i.e.,

$$\mathbf{P} = \mathbf{D} + \phi\,\mathbf{1}\mathbf{1}', \tag{3.2}$$

where $\mathbf{D}$ is a $v \times v$ diagonal matrix and ϕ is a scalar.

It is easy to verify that in a variance balanced design all the non-zero eigen values of **A** are equal to $v\,\phi$ and every normalized contrast is estimated with same variance $\sigma^2/v\,\phi$.

Remark 3.3: For an equireplicate, proper, binary, variance balanced design

$$\mathbf{N}\,\mathbf{N}' = (r - \lambda)\,\mathbf{I} + \lambda\mathbf{1}\mathbf{1}', \tag{3.3}$$

where λ is a scalar satisfying $\lambda\,(v - 1) = r\,(k - 1)$. Block designs with such characterization have been termed as Balanced Incomplete Block (BIB) designs and shall be studied in the following Section.

Theorem 3.2: For a BIB design the inequality $b \geqslant v$ holds.

Proof: From (3.3) we have

$$|\,\mathbf{N}\mathbf{N}'\,| = rk\,(r - \lambda)^{v-1} > 0.$$

Hence $R\,(\mathbf{N}\,\mathbf{N}') = v$. But the rank of **N** cannot exceed the number of columns in **N**. Hence

$$\begin{aligned} b &\geqslant R\,(\mathbf{N}) \\ &= R\,(\mathbf{N}\,\mathbf{N}') \\ &= v. \end{aligned}$$

This inequality is due to Fisher (1940). Atiquallah (1961) and Raghavarao (1962) proved the same result for binary, equi-replicate designs. An improved result, given in Theorem 3.3, was proved by Dey (1975).

Theorem 3.3: Barring trivial exceptions, the inequality $b \geqslant v$ holds for all connected, equireplicate, variance balanced designs.

Proof: For a balanced design, we have

$$\mathbf{A} = \theta\,(\mathbf{I} - \mathbf{1}\mathbf{1}'/v).$$

The non-zero eigen values of **A** are θ with multiplicity $v - 1$. Also, for an equi-replicate design,

$$\mathbf{A} = r\,\mathbf{I} - \mathbf{N}\mathbf{K}^{-1}\,\mathbf{N}'.$$

The roots of $\mathbf{N\,K^{-1}\,N'}$ are r and $r - \theta$ with multiplicity 1 and $v - 1$. Let $r - \theta > 0$. Then $R\,(\mathbf{N\,K^{-1}\,N'}) = R\,(\mathbf{N\,K^{-1/2}\,\,K^{-1/2}\,N'}) = R\,(\mathbf{N\,K^{-1/2}})$ $= v = R\,(\mathbf{N})$, as $\mathbf{K}$ is non-singular. But the rank of $\mathbf{N}$ cannot exceed the number of columns in $\mathbf{N}$. Hence

$$b \geqslant \text{Rank } \mathbf{N}$$

$$= v.$$

Let $r - \theta = 0$. Then $R\,(\mathbf{N\,K^{-1}\,N'}) = 1 = R\,(\mathbf{N})$. In view of $\mathbf{N\,1} = r\,\mathbf{1}$, the columns of $\mathbf{N}$ are proportional to $\mathbf{1}$, i.e. the design is orthogonal. Thus, barring those designs whose incidence matrices have identical rows, the inequality $b \geqslant v$ holds for all connected, equi-replicate, variance balanced designs.

3.2.2 Variance balanced designs

In Section 3.2.1 the concept of variance balance was introduced. We now study some important variance balanced designs.

The most important binary, equi-replicate and proper variance balanced designs are the BIB designs introduced by Yates (1936). In fact in the class of binary, equi-replicate and proper designs, BIB designs are the only variance balanced designs. A BIB design is an arrangement of v treatments or symbols in b blocks or sets of size k $(< v)$ satisfying the following conditions:

i) Every treatment occurs at most once in each block.
ii) Every treatment occurs in exactly r blocks.
iii) Every pair of treatments occurs together in λ blocks.

The parameters of a BIB design are v, b, r, k, λ. All these parameters are not independent. The following parametric relations hold:

$$vr = bk, \lambda\,(v-1) = r\,(k-1),$$

in addition to $b \geqslant v$, as proved in Theorem 3.2. All these conditions are, however, necessary only. We now prove some results on resolvable and affine resolvable BIB designs (for definition see Section 2.2).

Theorem 3.4: If a BIB design v, b, r, k, λ has the number of blocks b divisible by r, then Fisher's inequality can be refined to

$$b \geqslant v + r - 1. \tag{3.4}$$

Proof: From $vr = bk$ and $\lambda\,(v - 1) = r\,(k - 1$, we have

$$(b - r)/(v - 1) = (r - \lambda)/k$$

and

$$rk - v\lambda = r - \lambda.$$

Let $\quad b = nr, v = nk$, then

$$\frac{r-\lambda}{k} = \frac{rk - v\lambda}{k} = r - n\lambda.$$

Since $r > \lambda$, and $r - n\lambda$ is integral, we have

$$r - n\lambda \geqslant 1. \text{ Hence}$$

$$\frac{b-r}{v-1} \geqslant 1$$

or

$$b \geqslant v + r - 1.$$

It would, however, be of interest to study the conditions under which the equality in (3.4) holds.

For a resolvable BIB design it is possible to separate the b blocks into r sets $(S_0), (S_1), \ldots, (S_{r-1})$ with each set consisting of b/r blocks, so that every treatment occurs exactly once in the blocks of any set (S_i). Let the blocks belonging to the set (S_i) be $B_{i1}, B_{i2}, \ldots, B_{in}$ $(i = 0, 1, \ldots, r-1)$ where $b = nr$ and $v = nk$. Consider any particular block B_{01} of the set (S_0). Let s_{ij} be the number of treatments common to the block B_{01} and the block B_{ij} $i\,(\neq 0) = 1, 2, \ldots, r-1$; $j = 1, \ldots, n$. Let m denote the mean and s^2 the variance of $n\,(r-1)$ quantities s_{ij}.

We have

$$\sum_{i=1}^{r-1} \sum_{j=1}^{n} s_{ij} = k\,(r-1),$$

and

$$m = \frac{k}{n} = k^2/v.$$

Again the $k\,(k-1)/2$ pairs involved in B_{01} each appear $\lambda - 1$ times in the sets $(S_1), (S_2), \ldots, (S_{r-1})$. Hence

$\Sigma\Sigma\, s_{ij}\,(s_{ij} - 1) = (\lambda - 1)\, k\,(k-1)$, and $\Sigma\, s_{ij}^2 = k\,[(r-1) + (\lambda - 1)(k-1)]$.

Also $\quad \lambda = r\,(k-1)/(nk-1)$.

Hence

$$\Sigma\, s_{ij}^2 = \frac{k\,[(nk-1)\,(r-k) + r\,(k-1)^2]}{(nk-1)}$$

and

$$s^2 = \frac{\Sigma\,(s_{ij} - m)^2}{n\,(r-1)}$$

$$= \frac{k^2\,(n-1)\,[r\,(n-1) - (nk-1)]}{n^2\,(r-1)\,(nk-1)}$$

$$= \frac{k\,(v-k)\,(b - v - r + 1)}{n^2\,(r-1)\,(v-1)}$$

Since $s^2 > 0$, we have an alternative proof of the inequality $b \geqslant v + r - 1$.

If $b = v + r - 1$, then $s = 0$. Hence $s_{ij} = m$ for all values of i and j. This shows that the block B_{01} has exactly m treatments in common with each block of the sets $(S_1),\ldots, (S_{r-1})$. Since the choice of B_{01} was arbitrary, this shows that any two blocks of different sets have exactly m treatments in common, i.e. the design is affine resolvable. Also since m must be integral, k^2/v is integral.

Conversely, if the BIB design is affine resolvable, then s_{ij} must be constant for $i = 1, 2,\ldots, r-1$; $j = 1,\ldots,n$. Hence $s^2 = 0$ or $b = v + r - 1$. The constant value of s_{ij} is $m = k^2/v$.

Remark 3.4: The above result may sometimes be used to prove the non-resolvability of a BIB design even when b is divisible by r. Consider the BIB design $v = 6$, $b = 10$, $r = 5$, $k = 3$, $\lambda = 2$. If a resolvable solution could be obtained, it would be affine resolvable since $b = v + r - 1 = 10$. Then any two blocks belonging to different replications would have k^2/v treatments in common. This is impossible since k^2/v is non-integral. Hence no resolvable solution can be obtained.

A BIB design is said to be symmetric if $v = b$ and in consequence $r = k$. We now prove the following Theorem:

Theorem 3.5: For a symmetric BIB design $v = b$, $r = k$, λ, any two blocks have exactly λ treatments in common.

Proof: For a symmetric BIB design we have

$$\mathbf{N}\mathbf{1} = \mathbf{N}'\mathbf{1} = r\,\mathbf{1}.$$

But

$$\begin{aligned} \mathbf{N}\mathbf{N}'\mathbf{N} &= [(r-\lambda)\,\mathbf{I} + \lambda\,\mathbf{1}\mathbf{1}']\,\mathbf{N} \\ &= \mathbf{N}\,[(r-\lambda)\,\mathbf{I} + \lambda\,\mathbf{1}\mathbf{1}'] \\ &= \mathbf{N}\mathbf{N}\mathbf{N}' \end{aligned} \tag{3.5}$$

Also $|\,\mathbf{N}\mathbf{N}'\,| = r^2\,(r-\lambda)^{v-1} = |\,\mathbf{N}\,|^2 > 0$.

Hence $|\,\mathbf{N}\,| \neq 0$ which shows that $\mathbf{N}$ is non-singular. Premultiplying both sides of (3.5) by $\mathbf{N}^{-1}$, we have

$$\mathbf{N}'\mathbf{N} = \mathbf{N}\mathbf{N}'.$$

If $j \neq \beta$ $(1 \leqslant j \leqslant v;\ 1 \leqslant \beta \leqslant b)$, then the element in the j-th row and β-th column of $\mathbf{N}'\mathbf{N}$ is the same as the corresponding element in $\mathbf{N}\mathbf{N}'$ and is therefore λ. This is also equal to the inner product of j-th and β-th column of $\mathbf{N}$. But this is the number of treatments common in the j-th and β-th block of a symmetric BIB design.

Given a BIB design $D(v, b, r, k, \lambda)$, another BIB design can be obtained by interchanging 0 and 1 in the incidence matrix of the design D. The resulting design D^*, called the complementary design, has the parameters $v^* = v$, $b^* = b$, $r^* = b - r$, $k^* = v - k$, $\lambda^* = b - 2r + \lambda$. The property of symmetrical BIB design given in Theorem 3.5 can be exploited to obtain two more BIB designs. If we delete one of the blocks of D ($v = b$, $r = k$, λ) and delete all the treatments of the deleted block from the remaining blocks of D, then the resulting design D_1, called the residual design, is a BIB design with parameters $v_1 = v - k$, $b_1 = b - 1$, $r_1 = r$, $k_1 = k - \lambda$, $\lambda_1 = \lambda$. The process by which the new design D_1 is obtained is called the process of block section. Further, if we obtain another design from the remaining $v - 1$ blocks of D by retaining only those treatments in these blocks which are contained in the deleted block, then the resulting design D_2, called the derived design, is again a BIB design with parameters $v_2 = k$, $b_2 = b - 1$, $r_2 = r - 1$, $k_2 = \lambda$, $\lambda_2 = \lambda - 1$. The process by which the new design D_2 is obtained is called the process of block intersection. The converse of the above two methods, namely that of constructing a BIB design from the residual and derived designs by embedding them has been of great interest to combinatorialists. It was shown by Bhattacharya (1945) that the converse is not true. It is also known that the dual of a symmetrical BIB design is also a BIB design with the same parameters as of the original design.

BIB designs are very useful block designs and considerable amount of work is available in the literature on construction and combinatorial properties of these designs. Interested readers may refer to the texts of Raghavarao (1971), Hall (1967) and Street and Street (1987).

In a BIB design every pair of treatments occurs together in λ blocks. This idea of constancy of pairwise occurrence of treatments in a BIB design can be extended to the constancy of every triplet, or, in general, every t-plet of the treatments in the design. Such designs are called t-designs and the 3-designs are called doubly balanced designs (Calvin, 1954). For a detailed discussion on t-designs, the readers may refer to Hedayat and Kageyama (1980) and Kageyama and Hedayat (1983).

BIB designs have recently been used in sampling theory as well. For an interesting application of BIB designs (2- and 3-designs) the readers may refer to Foody and Hedayat (1977), Wynn (1977) and Gupta, Nigam and Kumar (1982).

So far the discussion was restricted to BIB designs. In BIB designs the block size is less than the number of treatments in the design. However, if one considers p-ary balanced block designs then the block size need not necessarily be smaller than the number of treatments in the design. Tocher (1952) obtained some balanced ternary designs by trial and error. Das and Rao (1968), Murty and Das (1967), Dey (1970), Saha and Dey (1973), Nigam (1974), Saha (1975), Kageyama (1976), Morgan (1977) and Kageyama and Tsuji (1980) studied the p-ary balanced designs in detail and advanced several methods of constructing proper and equi-replicate designs.

Rao (1958) showed that if a binary, balanced design is proper, then it must be equi-replicate and gave an example of an equi-replicated, binary, balanced design with unequal block sizes. John (1964) derived examples showing that the result of Rao (1958) does not hold when either of the requirements of binary or equal block sizes is relaxed. He gave some examples of balanced, binary and non-binary designs with varying replications.

Kulshreshtha, Dey and Saha (1972) and Hedayat and Federer (1974) presented methods of construction of balanced, binary and ternary designs with two unequal block sizes and varying replications. Their methods of construction involved the union of blocks of two block designs and augmentation of some blocks with additional treatments. Puri and Nigam (1976) generalized their methods to obtain balanced p-ary designs in $(v + 1)$ treatments by unionization of two balanced p-ary designs.

Tyagi (1979) gave a method of construction of balanced designs with unequal block sizes and varying replications and unequal concurrences λ. An interesting property of these designs is that these require less number of experimental units as compared to the corresponding BIB designs with same precision. These designs were constructed by utilizing the unreduced BIB designs with $k = 2$. Khatri (1982) extended the method further to construct designs with $k = 3$. For some further work on balanced p-ary designs the readers may refer to Shafiq and Federer (1979).

We now discuss partially balanced incomplete block (PBIB) designs. Discussions on several other useful partially balanced designs like cyclic and supplemented block designs is deferred to later Chapters.

3.3 PARTIALLY BALANCED INCOMPLETE BLOCK DESIGNS

PBIB designs were introduced by Bose and Nair (1939) as a generalization of BIB designs. In these designs all the elementary contrasts are not estimated with the same variance. Bose and Shimamoto (1952) introduced the concept of partially balanced association scheme and the definition of PBIB designs was rephrased in this context. The definitions of m-class association scheme and the PBIB design are given below.

Definition 3.2: Given v treatments 1, 2, ..., v, a relation satisfying the following conditions is said to be an association scheme with m classes:

i) Any two treatments are either 1st, 2nd, ..., or m-th associates and the relation of association is symmetrical. In other words, if the treatment α is i-th associate of the treatment β, then β is also the i-th associate of α.

ii) Each treatment α has n_i i-th associates and the number n_i is independent of the particular treatment α chosen.

iii) If any two treatments α and β are i-th associates, then the number of treatments that are j-th associate of α and k-th associate of β is p^i_{jk} and the number p^i_{jk} is independent of the pair of i-th associate α and β.

The numbers v, n_i $(i = 1, ..., m)$ and p^i_{jk} $(i, j, k = 1, ..., m)$ are called the parameters of the association scheme. The parameters p^i_{jk} can be written in the form of m matrices $\mathbf{P}_i = (p^i_{jk})$, $i = 1, ..., m$.

Definition 3.3: Given an m-associate scheme, a PBIB design with m-associate classes, based on the association scheme, is an arrangement of v-treatments into b blocks such that

i) each block contains k distinct treatments $(k < v)$,
ii) each treatment occurs in r blocks,
iii) any two treatments which are i-th associates occur together in λ_i blocks $(i = 1, ..., m)$, where λ_i is a constant independent of the pair of the i-th associates chosen.

The symbols v, b, r, k, λ_i $(i = 1, ..., m)$ are called the parameters of the design. The concept of the association scheme plays a fundamental role in the analysis and classification of PBIB designs. Nair and Rao (1942) have shown that λ_i parameters need not be distinct. Bose and Mesner (1959) showed that the condition $p^i_{jk} = p^i_{kj}$ in the earlier definition of PBIB designs is redundant.

For convenience, assume that every treatment is 0-th associate of itself and of no other treatment. Clearly $n_0 = 1$, $p^0_{ij} = n_i \delta_{ij}$, $p^i_{0k} = \delta_{ik}$, $\lambda_0 = r$ $(i, j, k = 1, ..., m)$ shows that the parameters of PBIB design and its association scheme satisfy the following parametric relations

$$vr = bk, \sum_{i=1}^{m} n_i = v - 1, \sum_{i=0}^{m} n_i \lambda_i = rk,$$

$$\sum_{k=0}^{m} p^i_{jk} = n_j, \text{ and } n_i p^i_{jk} = n_j p^j_{ik},$$

where δ_{ij} is the Kronecker delta, taking the value 1 if $i = j$, and 0 otherwise.

The concept of association matrices was introduced by Bose and Mesner (1959). Let us define $v \times v$ matrices

$$\mathbf{B}_i = (b^i_{\alpha\beta}),\ i = 1, 2, ..., m,$$

where $b^i_{\alpha\beta}$ is the (α, β)-th element of $\mathbf{B}_i$ satisfying

$$\begin{aligned} b^i_{\alpha\beta} &= 1 \text{ if } \alpha \text{ and } \beta \text{ are } i\text{-th associates,} \\ &= 0 \text{ otherwise.} \end{aligned}$$

The $\mathbf{B}_i$ matrices, called the association matrices of the association scheme, are symmetric with row and column totals each equal to n_i, $i = 1, ..., m$. These matrices have unity in the mutually exclusive positions on each row because a given treatment is either a first associate or a second associate, ..., or m-th associate of any other treatment. It, therefore, follows that $\mathbf{B}_0 = \mathbf{I}_v$, and $\sum_{i=0}^{m} \mathbf{B}_i = \mathbf{1}\,\mathbf{1}'$.

Further, it can be shown easily that

$$\sum_{i=0}^{m} c_i \mathbf{B}_i = \mathbf{0}$$

holds if and only if $c_i = 0$ for all i, showing thereby that the linear functions of $\mathbf{B}_0, \mathbf{B}_1, \ldots, \mathbf{B}_m$ form a vector space of $(m+1)$ dimensionality with basis $\mathbf{B}_0, \mathbf{B}_1, \ldots, \mathbf{B}_m$.

Since the (α, β)-th element of $\mathbf{B}_j \mathbf{B}_k$ can be interpreted as the number of treatments common to the j-th associates of α and k-th associates of β, we have

$$\mathbf{B}_j \mathbf{B}_k = \sum_{i=0}^{m} p^i_{jk} \mathbf{B}_i = \sum_{i=0}^{m} p^i_{kj} \mathbf{B}_i = \mathbf{B}_k \mathbf{B}_j,\ j, k = 0, 1, \ldots, m.$$

Also if $\mathbf{N}$ is the incidence matrix of a PBIB design then Connor and Clatworthy (1954) and also Bose and Mesner (1959) showed that the distinct eigen values of

$$\mathbf{N}\,\mathbf{N}' = r\,\mathbf{B}_0 + \lambda_1 \mathbf{B}_1 + \ldots + \lambda_m \mathbf{B}_m \qquad (3.6)$$

and

$$\mathbf{P}^* = r\,\mathbf{P}^*_0 + \lambda_1 \mathbf{P}^*_1 + \ldots + \lambda_m \mathbf{P}^*_m, \qquad (3.7)$$

are same. Here $\mathbf{P}^*_i = (p^i_{jk}),\ i, j, k = 0, 1, \ldots, m$ are matrices of order $m + 1$.

The association matrices play a fundamental role in the study of association schemes. Using the properties of association matrices, Bose and Mesner (1959) obtained many interesting results concerning the combinatorial properties of association schemes. Kusumoto (1967) has given a necessary and sufficient condition under which $(m+1)$ symmetric matrices are the association matrices of an association scheme with m-classes.

From practical view point, the two-associate class PBIB designs are important and an exhaustive list of these designs was prepared by Bose, Clatworthy and Shrikhande (1954). The list was subsequently revised by Clatworthy (1956, 1973). Considerable amount of work has been done by various workers to study two class association schemes and related designs. Literature on higher associate class is very widespread and it is beyond the scope of this book to present an account of the same. Their utility is also limited because their analysis becomes considerably involved with an increase in the number of associate classes. We restrict ourselves to giving a brief exposition of some of the widely used association schemes and defer to Chapter 4 a detailed discussion on designs based upon Kronecker product association schemes.

3.3.1 Two-associate class PBIB designs

The first attempt to classify two associate PBIB designs was made by Bose and Shimamoto (1952).

Before we study two associate class **PBIB** schemes and designs, it may be useful to describe the work of Connor and Clatworthy (1954). They determined the eigen values of **NN′** and their multiplicities for any **PBIB** design with two-associate classes. The three distinct eigen values are

$$\theta_0 = rk$$

and

$$\theta_i = r - \tfrac{1}{2}[(\lambda_1 - \lambda_2)\{-\gamma + (-1)^i\sqrt{\Delta}\} + (\lambda_1 + \lambda_2)],\ i = 1,\ 2, \quad (3.8)$$

where

$$\gamma = p^2_{12} - p^1_{12},\ \Delta = \gamma^2 + 2\beta + 1,\ \beta = p^1_{12} + p^2_{12}. \quad (3.9)$$

The eigen values have the multiplicities $\rho_0 = 1$, ρ_1, ρ_2 such that $\rho_1 + \rho_2 = v - 1$. Again, noting that trace of the matrix **NN′** is vr, we have

$$rk + \rho_1\theta_1 + \rho_2\theta_2 = vr \quad (3.10)$$

which on simplification, subject to $\rho_1 + \rho_2 = v - 1$, gives

$$\rho_i = \frac{n_1 + n_2}{2} + (-1)^i\left[\frac{(n_1 - n_2) + \gamma(n_1 + n_2)}{2\sqrt{\Delta}}\right]. \quad (3.11)$$

The multiplicities therefore depend only on parameters of the association scheme and are independent of the design parameters. Since multiplicities must be integral, this provides a necessary condition for an association scheme to exist.

3.3.1.1 Group divisible designs

A PBIB design with two associate classes is group divisible (G.D.) if $v = mn$ treatments satisfy the following association scheme: the v treatments are divided into m groups each of size n such that any two treatments of the same group are first associates and that of different groups are second associates. The parameters of the scheme are $n_1 = n - 1$, $n_2 = n(m-1)$ and

$$\mathbf{P}_1 = (p^1_{jk}) = \begin{bmatrix} n-2 & 0 \\ 0 & n(m-1) \end{bmatrix}, \quad (3.12)$$

$$\mathbf{P}_2 = (p^2_{jk}) = \begin{bmatrix} 0 & n-1 \\ n-1 & n(m-2) \end{bmatrix}.$$

Using the expressions (3.8) and (3.11), the eigen values of **NN′** simplify to

$$\theta_0 = rk,\ \theta_1 = r - \lambda_1,\ \theta_2 = rk - v\lambda_2,\ \rho_0 = 1,\ \rho_1 = m(n-1)$$

and $\rho_2 = m - 1$. Bose and Connor (1952) further classified these into (i) singular(S) if $\theta_1 = 0$, (ii) semi-regular (SR) if $\theta_1 > 0$, $\theta_2 = 0$ and (iii) regular (R) if $\theta_1, \theta_2 > 0$.

3.3.1.2 Triangular designs

A PBIB design with two associate classes is triangular if $v = n(n-1)/2$ treatments satisfy the following association scheme: the v treatments are arranged in an array of n rows and n columns such that the principal diagonal (upper left hand to the lower right hand corner) is left blank and $n(n-1)/2$ positions in each of the triangles above and below the diagonals are filled by $n(n-1)/2$ treatments in a symmetric way and the array is symmetrical about the principal diagonal. The treatments in the same row and column are first associates and others are second associates. For instance, when $n = 5$, the ten treatments can be arranged as follows:—

*	1	2	3	4
1	*	5	6	7
2	5	*	8	9
3	6	8	*	10
4	7	9	10	*

The first associates of 1 are 2, 3, 4, 5, 6, 7 and second associates of 1 are 8, 9, 10.

The following results hold for such an association scheme:

$$n_1 = 2(n-2),\ n_2 = (n-2)(n-3)/2,$$

$$\mathbf{P}_1 = \begin{bmatrix} n-2 & n-3 \\ n-3 & (n-3)(n-4)/2 \end{bmatrix} \tag{3.13}$$

$$\text{and } \mathbf{P}_2 = \begin{bmatrix} 4 & 2n-8 \\ 2n-8 & (n-4)(n-5)/2 \end{bmatrix},$$

where $n \geqslant 5$. When $n = 4$, $p^2_{12} = 0$ and the design is also GD.

An alternate way of defining the triangular association scheme was proposed by John (1966). Let any pair of treatments be represented by the ordered pair of integers (i_1, i_2), $0 < i_1 < i_2 \leqslant n$. Then two treatments are first associates if their representations have an integer in common, otherwise, they are second associates. The eigen values of $\mathbf{NN}'$ other than rk are

$$r + (n-4)\lambda_1 - (n-3)\lambda_2 \text{ and } (r - 2\lambda_1 + \lambda_2)$$

with multiplicities $(n-1)$ and $n(n-3)/2$ respectively.

3.3.1.3 Latin square type designs ($\mathbf{L}_i$)

A two associate PBIB design is a Latin Square type with i constraints if

$v = n^2$ treatments have the following association scheme: the v treatments are arranged in a square array of order n and the array is superimposed on $i-2$ mutually orthogonal latin squares of side n. Two treatments are first associates if and only if they appear in the same row or column of the array or correspond to the same letter in any of the $(i-2)$ latin squares. For L_2 scheme two treatments are first associates if they occur in the same row or column. For this scheme we have

$$n_1 = i(n-1),\ n_2 = (n-i+1)(n-1),$$

$$\mathbf{P}_1 = \begin{bmatrix} i^2 - 3i + n & (i-1)(n-i+1) \\ (i-1)(n-i+1) & (n-i)(n-i+1) \end{bmatrix} \tag{3.14}$$

and
$$\mathbf{P}_2 = \begin{bmatrix} i(i-1) & i(n-i) \\ i(n-i) & (n-i)^2 + (i-2) \end{bmatrix}$$

The eigen values of $\mathbf{NN}'$ other than rk are $r + (n-i)\lambda_1 - (n-i+1)\lambda_2$ and $r - i\lambda_1 + (i-1)\lambda_2$ with the respective multiplicities $i(n-1)$ and $(n-i+1)(n-1)$.

A simple lattice is a PBIB with two associate classes based on L_2 association scheme. Similarly, the triple lattice is a two associate PBIB with L_3 association scheme. Simple and triple lattice belong to the wider class of lattice designs introduced by Yates (1937). These were also called quasifactorial designs. The reader may refer to the texts of Kempthorne (1952) and Federer (1955) for a detailed exposition of lattice designs.

We have so far discussed the three most useful classes of two-associate PBIB designs classified by Bose and Shimamoto (1952). The fourth class is not so important and is based upon the cyclic association scheme. In view of its restricted usefulness, we avoid a discussion on the same. Yet another class of two associate class PBIB designs is the one for which either (i) $\lambda_1 \neq 0, \lambda_2 = 0$ or (ii) $\lambda_1 = 0, \lambda_2 \neq 0$. Such a class is termed 'simple' and many of the designs based upon G.D., triangular, etc. may also satisfy (i) or (ii). These designs, however, have a close connection with two associate PBIB designs based upon partial geometry (r, k, t) introduced by Bose (1963).

3.3.1.4 Partial geometry (r, k, t)

Bose (1963) defined a partial geometry (r, k, t) as a system of v points and b lines with their incidence relation satisfying the following axioms:

Axiom 1. Any two distinct points are incident with not more than one line.

Axiom 2. Each point is incident with r lines.
Axiom 3. Each line is incident with k points.
Axiom 4. If the point P is not incident on p, then there are exactly t lines through P that intersect p.

Identifying v points with v treatments and b lines with b blocks it is easily seen that every partial geometry is isomorphic to a simple PBIB design with two associate classes and with the parameters

$$v = k\,[(r-1)(k-1)+t]/t,\ b = r\,[(r-1)(k-1)+t]/t,$$

$$r, k, \lambda_1 = 1, \lambda_2 = 0, n_1 = r\,(k-1), n_2 = (r-1)(k-1)(k-t)/t,$$

$$p^1_{11} = (t-1)(r-1)+k-2, p^1_{12} = (r-1)(k-1), p^2_{11} = rt,$$

$$p^2_{12} = r(k-t-1),\ 1 \leqslant t \leqslant \min(r, k). \tag{3.15}$$

For the above association scheme based on partial geometry (r, k, t), any two points are first associates which are incident with a line of the geometry and those two are second associates which are not incident with a line. The eigen values of $\mathbf{NN}'$ matrix are $\theta_0 = rk$, $\theta_1 = r+k-t-1$ and $\theta_2 = 0$ with respective multiplicities $\rho_0 = 1$, $\rho_1 = [rk\,(r-1)(k-1)]/t\,(k+r-t-1)$ and $\rho_2 = v-1-\rho_1$. Since the multiplicities of eigen values need to be integral, it follows that necessary condition for the existence of a partial geometry (r, k, t) is that $rk\,(r-1)(k-1)/t(k+r-t-1)$ is a positive integer.

3.4 EFFICIENCY BALANCE

In Section 3.2.2 we studied variance balanced designs. These designs are balanced because every normalised contrast of treatments is estimated through the design with same precision. However, the efficiency of every normalised contrast may not be same. The efficiency of any treatment contrast is defined as the ratio of the variance of the estimate of a treatment contrast to the variance of the estimate of the same treatment contrast through an orthogonal design with the same vector of replications and same variance per plot (Pearce, 1968, 1970). In this Section we study another class of designs which permit estimation of some special contrasts of treatment effects with same efficiency. This subject has been studied by Kshirsagar (1958), Jones (1959), Calinski (1971), Puri and Nigam (1975b), Williams (1975), Pal (1980). The results reported here are due to Dey and Gupta (1986). We first give some definitions and Lemmas:

Lemma 3.1: Let $\mathbf{A}$ be any real square matrix of order n having eigen values μ_t, $t = 1,\ldots,s$, with respective multiplicities ρ_t, $\Sigma\rho_t = n$, and n right eigen vectors $\mathbf{u}_{tj}$, $j = 1,\ldots,\rho_t$, i.e., $\mathbf{Au}_{tj} = \mu_t\mathbf{u}_{tj}$. Then n eigen vectors $[\mathbf{u}_{tj}]$ are linearly independent if and only if $R\,(\mathbf{A}-\mu_t\mathbf{I}) = n-\rho_t$ for all $t = 1,\ldots,s$. Further if $\mathbf{U} = [\mathbf{u}_{11},\ldots, \mathbf{u}_{1\rho_1},\ldots, \mathbf{u}_{s1},\ldots, u_{s\rho_s}]$ then $\mathbf{U}^{-1}\mathbf{AU} = \text{diag}\,(\mu_1 \ldots \mu_1 \ldots \mu_s \ldots \mu_s)$.

For a proof of Lemma 3.1, see Searle (1982, p. 284).

Remark 3.5: The n eigen vectors $[\mathbf{u}_{ij}]$ are linearly independent but not mutually orthogonal. However, if $\mathbf{A}$ is symmetric, then we can always have a set of mutually orthogonal eigen vectors.

Let $\mu_1, \ldots, \mu_s$ be the distinct eigen values of matrix $\mathbf{M}$, other than simple eigen value unity, with multiplicities $\rho_1, \ldots, \rho_s$, $\sum_{i=1}^{s} \rho_i = v-1$, $0 < \mu_i < 1$ for all $i = 1, \ldots, s$ (see Section 2.5). Then we have the following Lemma:

Lemma 3.2: $R(\mathbf{M} - \mu_i \mathbf{I}) = v - \rho_i$ for all $i = 1, \ldots, s$.

Proof: From (2.10) we have

$$\mathbf{M} - \mu_i \mathbf{I} = \mathbf{R}^{-1/2} (\mathbf{B} - \mu_i \mathbf{I}) \mathbf{R}^{1/2}$$

showing thereby that $\mathbf{M} - \mu_i \mathbf{I}$ is similar to $\mathbf{B} - \mu_i \mathbf{I}$ and thus $R(\mathbf{M} - \mu_i \mathbf{I}) = R(\mathbf{B} - \mu_i \mathbf{I}) = v - \rho_i$, $i = 1, \ldots, s$.

Let $\mu_0, \mu_1, \ldots, \mu_s$ be the distinct eigen values of the matrix $\mathbf{B}$ given by

$$\mathbf{B} = \mathbf{R}^{-1/2} \mathbf{N} \mathbf{K}^{-1} \mathbf{N}' \mathbf{R}^{-1/2},$$

with respective multiplicities $\rho_0 = 1, \rho_1, \ldots, \rho_s$. Let the corresponding orthonormalised eigen vectors be $\mathbf{p}_{ij}$, $i = 0, 1, \ldots, s$; $j = 1, \ldots, \rho_i$. Then the matrix $\mathbf{B}$ has the spectral decomposition

$$\mathbf{B} = \sum_{i=0}^{s} \mu_i \sum_{j=1}^{\rho_i} \mathbf{p}_{ij} \mathbf{p}'_{ij}.$$

In view of Lemma 2.7, we thus have

$$\begin{aligned} \mathbf{M} &= \mathbf{R}^{-1/2} \mathbf{B} R^{1/2} \\ &= \sum_{i=0}^{s} \mu_i \sum_{j=1}^{\rho_i} \mathbf{R}^{-1/2} \mathbf{p}_{ij} \mathbf{p}'_{ij} \mathbf{R}^{1/2} \\ &= \sum_{i=0}^{s} \mu_i \sum_{j=1}^{\rho_i} \mathbf{a}_{ij} \mathbf{b}'_{ij}, \qquad (3.16) \\ &= \sum_{i=0}^{s} \mu_i \sum_{j=1}^{\rho_i} \mathbf{a}_{ij} \mathbf{a}'_{ij} \mathbf{R} \end{aligned}$$

where $\mathbf{a}_{ij} = \mathbf{R}^{-1/2} \mathbf{p}_{ij}$ and $\mathbf{b}_{ij} = \mathbf{R}^{1/2} \mathbf{p}_{ij}$ are respectively the right and left eigen vectors of $\mathbf{M}$ corresponding to the eigen value μ_i. In view of remark 3.5 the eigen vectors $[\mathbf{a}_{ij}]$ (and so also $[\mathbf{b}_{ij}]$) are linearly independent but not mutually orthogonal unless the design is equireplicate.

$$\text{Let } \sum_{j=1}^{\rho_i} \mathbf{a}_{ij} \mathbf{b}'_{ij} = \mathbf{L}_i = \sum_{j=1}^{\rho_i} \mathbf{a}_{ij} \mathbf{a}'_{ij} \mathbf{R} \text{ for } i = 0, 1, \ldots s. \qquad (3.17)$$

Then we have

$$\mathbf{M} = \sum_{i=0}^{s} \mu_i \mathbf{L}_i \qquad (3.18)$$

Note that $\mathbf{L}_i^2 = \mathbf{L}_i$ and $\mathbf{L}_i \mathbf{L}_m = \mathbf{0}$, if $i \neq m$. (3.19)

Also $R(\mathbf{L}_i) = \rho_i$ and $\sum_{i=0}^{s} \mathbf{L}_i = \mathbf{I}$. (3.20)

Thus, $\mathbf{L}_i$ are mutually orthogonal idempotent matrices. The representation of **M** in terms of the eigen values and mutually orthogonal idempotent matrices was first given by Calinski (1971). Calinski stated such a representation of **M** is possible for those block designs for which **M** is semi-simple. However, **M** is semi-simple for any connected block design (vide Lemmas 3.1 and 3.2). Summarizing, we thus have the following result:

Theorem 3.6: Let $\mu_0 = 1, \mu_1, \ldots, \mu_s$ be the eigen values of **M** with respective multiplicities $\rho_0 = 1, \rho_1 \ldots, \rho_s$, then

$$\mathbf{M} = \sum_{i=0}^{s} \mu_i \mathbf{L}_i,$$

where $\mathbf{L}_i$'s are mutually orthogonal idempotent matrices with $R(\mathbf{L}_i) = \rho_i$ and

$$\sum_{i=0}^{s} \mathbf{L}_i = \mathbf{I}.$$

Corollary 3.2: The matrix $\mathbf{M}_0 = \mathbf{M} - \mathbf{1}\,\mathbf{r}'/n$ has the representation

$$\mathbf{M}_0 = \sum_{i=1}^{s} \mu_i \mathbf{L}_i, \tag{3.21}$$

where $R(\mathbf{L}_i) = \rho_i$, $i = 1, \ldots, s$, and $\sum_{i=1}^{s} \mathbf{L}_i = \mathbf{I} - \mathbf{1}\,\mathbf{r}'/n$.

Consider now a block design $D(v, b, \mathbf{r}, \mathbf{k})$. Let $\mathbf{s}'\,\mathbf{R}\,\boldsymbol{\tau}$ be a contrast of treatment effects, i.e., $\mathbf{s}'\,\mathbf{R}\,\mathbf{1} = \mathbf{s}'\,\mathbf{r} = 0$. Under the fixed effects model (2.2), the best linear unbiased estimator of $\mathbf{s}'\,\mathbf{R}\boldsymbol{\tau}$ is $\mathbf{s}'\,\mathbf{R}\,\hat{\boldsymbol{\tau}}$, where $\hat{\boldsymbol{\tau}}$ is a solution of the equations (2.6). Thus

$$\begin{aligned} \mathbf{s}'\,\mathbf{R}\hat{\boldsymbol{\tau}} &= \mathbf{s}'\,\mathbf{R}\,\mathbf{A}^{-}\,\mathbf{Q} \\ &= \mathbf{s}'\mathbf{R}\,\boldsymbol{\Omega}\,\mathbf{Q}. \end{aligned} \tag{3.22}$$

But

$$\boldsymbol{\Omega} = (\mathbf{I} - \mathbf{M}_0)^{-1}\,\mathbf{R}^{-1},$$

so that

$$\begin{aligned} \mathbf{s}'\,\mathbf{R}\,\boldsymbol{\Omega} &= \mathbf{s}'\mathbf{R}\,(\mathbf{I} - \mathbf{M}_0)^{-1}\,\mathbf{R}^{-1} \\ &= \mathbf{s}'\mathbf{R}\,(\mathbf{R} - \mathbf{R}\,\mathbf{M}_0)^{-1} \\ &= \mathbf{s}'\mathbf{R}\,(\mathbf{R} - \mathbf{M}_0'\,\mathbf{R})^{-1} \text{ since } \mathbf{R}\mathbf{M}_0 \text{ is symmetric,} \\ &= \mathbf{s}'\mathbf{R}\,\mathbf{R}^{-1}\,(\mathbf{I} - \mathbf{M}_0')^{-1} \end{aligned}$$

$$= \mathbf{s}'(\mathbf{I} - \mathbf{M}_0')^{-1} \tag{3.23}$$

$$= \mathbf{s}'(\mathbf{I} + \sum_{h=1}^{\infty} \mathbf{M}_0'^h),$$

for, $\mathbf{M}_0$ and $\mathbf{M}_0'$ both have the same set of eigen values.

Now suppose that $\mathbf{s}$ is a right eigen vector of $\mathbf{M}_0$ corresponding to the eigen value μ, $0 \leqslant \mu < 1$, i.e.

$$\mathbf{M}_0\mathbf{s} = \mu\mathbf{s}. \tag{3.24}$$

Then, clearly,

$$\mathbf{s}'\mathbf{M}_0'^h = \mu^h\mathbf{s}' \text{ for } h = 1, 2, \ldots,$$

and we have from (3.13) and (3.14)

$$\mathbf{s}'\mathbf{R}\hat{\boldsymbol{\tau}} = (\mathbf{s}' + \mathbf{s}' \sum_{h=1}^{\infty} \mu^h)\,\mathbf{Q}$$

$$= [1 + \mu/(1-\mu)]\,\mathbf{s}'\mathbf{Q}$$

$$= \mathbf{s}'\mathbf{Q}/(1-\mu). \tag{3.25}$$

Thus, the best linear unbiased estimator of $\mathbf{s}'\mathbf{R}\boldsymbol{\tau}$, under condition (3.24), is simply $\mathbf{s}'\mathbf{Q}/(1-\mu)$.

The variance of this estimator using the design D is

$$V(\mathbf{s}'\mathbf{R}\hat{\boldsymbol{\tau}}) = (1-\mu)^{-2}\, V(\mathbf{s}'\mathbf{Q})$$

$$= (1-\mu)^{-2}\,\mathbf{s}'\mathbf{A}\mathbf{s}\sigma^2 \tag{3.26}$$

But

$$\mathbf{A} = \mathbf{R}(\mathbf{I} - \mathbf{M}) = \mathbf{R}(\mathbf{I} - \mathbf{M}_0 - \mathbf{1}\mathbf{r}'/n)$$

$$= (\mathbf{I} - \mathbf{M}_0' - \mathbf{r}\mathbf{1}'/n)\,\mathbf{R},$$

because $\mathbf{A}$ is symmetric. Thus

$$\mathbf{s}'\mathbf{A}\mathbf{s} = \mathbf{s}'\,(\mathbf{I} - \mathbf{M}_0' - \mathbf{r}\mathbf{1}'/n)\;\mathbf{R}\mathbf{s}$$

$$= \mathbf{s}'\mathbf{R}\mathbf{s} - \mathbf{s}'\mathbf{M}_0'\mathbf{R}\mathbf{s} \tag{3.27}$$

as $\mathbf{s}'\mathbf{r} = 0$.

Using (3 24) in (3.27), we have

$$\mathbf{s}'\mathbf{A}\mathbf{s} = \mathbf{s}'\mathbf{R}\mathbf{s} - \mu\mathbf{s}'\mathbf{R}\mathbf{s}$$

$$= (1-\mu)\,\mathbf{s}'\mathbf{R}\mathbf{s}. \tag{3.28}$$

Thus,

$$V(\mathbf{s}'\mathbf{R}\hat{\boldsymbol{\tau}}) = \{\mathbf{s}'\mathbf{R}\mathbf{s}/(1-\mu)\}\;\sigma^2, \tag{3.29}$$

It may be noted that when $\mathbf{s}$ is a right eigen vector of $\mathbf{M}_0$ corresponding to the eigen value μ, then $\mathbf{s}'\mathbf{R}\boldsymbol{\tau}$ is a contrast because of $\mathbf{s}'\mathbf{r} = 0$. This is so (because $\mathbf{r}^{1/2}$ is an eigen vector of $\mathbf{B}$ corresponding to the eigen value unity see 2.11) and therefore all other eigen vectors of $\mathbf{B}$ are orthogonal to $\mathbf{r}^{1/2}$.

Suppose 0 is an orthogonal design (possibly hypothetical) relative to design D, with the same replication vector $\mathbf{r}$. Then the best linear unbiased estimator of $\mathbf{s}'\mathbf{R}\boldsymbol{\tau}$, estimated through the design 0, is $\mathbf{s}'\mathbf{Q}$ and this estimator has variance

$$V\,(\mathbf{s}'\mathbf{R}\hat{\boldsymbol{\tau}})_0 = \sigma^2\,\mathbf{s}'\mathbf{R}\mathbf{s}. \tag{3.30}$$

Thus the efficiency of the contrast $\mathbf{s}'\mathbf{R}\hat{\boldsymbol{\tau}}$ is

$$E = \mathbf{s}'\mathbf{R}\mathbf{s}/\mathbf{s}'\mathbf{R}\mathbf{s}\,(1-\mu)^{-1} = (1-\mu). \tag{3.31}$$

From the foregoing, it is therefore clear that if $\mathbf{M}$ has eigen values 1, $\mu_1, \mu_2, \ldots, \mu_s$ with respective multiplicities $1, \rho_1, \rho_2, .., \rho_s$ and if $[\mathbf{s}_{ij}]$, $j=1,\ldots, \rho_i$ are the right eigen vectors of $\mathbf{M}_0$ corresponding to the eigen value μ_i, $0 \leqslant \mu_i < 1$ for $i = 1, .., s$, then the contrasts $[\mathbf{s}'_{ij}\,\mathbf{R}\boldsymbol{\tau}]$ (ρ_i in number) are each estimated with an efficiency factor $(1-\mu_i)$, for $i = 1, \ldots, s$. Such designs have been termed partially efficiency balanced with s efficiency classes, by Puri and Nigam (1977a). The contrasts $[\mathbf{s}_{ij}'\mathbf{R}\boldsymbol{\tau}]$ or simply $\mathbf{s}_{ij}$ have been termed as basic contrasts by Pearce, Calinski and Marshall (1974).

From the analysis presented above it is apparent that every conceivable connected block design is a partially-efficiency balanced design with s efficiency classes for some integer s, $1 \leqslant s \leqslant v-1$. Thus a general partially efficiency balanced design is not of much interest and the concept itself is rather weak. We therefore take up some special cases which could be of interest. First we study the case when $s = 1$. When $s = 1$, i.e. all μ_i's are equal ($1 \leqslant i \leqslant s$), say μ, the design is termed as Efficiency-balanced (Puri and Nigam, 1975b; Williams, 1975). We thus have

Definition 3.4: A design $D(v, b, \mathbf{r}, \mathbf{k})$ is efficiency balanced if all the $(v-1)$ treatment contrasts $[\mathbf{s}'\mathbf{R}\boldsymbol{\tau}]$ such that $\mathbf{M}_0\mathbf{s} = \mu\mathbf{s}$ are estimated with the same efficiency μ, where μ is the unique non-zero eigen value of $\mathbf{M}_0$ with multiplicity $(v-1)$.

From (3.21) it follows that the $\mathbf{M}_0$ matrix of an efficiency-balanced design is of the form

$$\mathbf{M}_0 = \mu\,(\mathbf{I} - \mathbf{1}\mathbf{r}'/n). \tag{3.32}$$

The $\boldsymbol{\Omega}$-matrix of efficiency-balanced designs, using (2.16), is

$$\boldsymbol{\Omega} = \mu/(1-\mu)\left[\frac{1}{\mu}\mathbf{I} - \mathbf{1}\mathbf{r}'/n\right].$$

Moreover, since

$$\mathbf{P} = \mathbf{R}\mathbf{M} = \mathbf{R}\mathbf{M}_0 + \mathbf{r}\mathbf{r}'/n,$$

in view of (3.32) we have

$$\mathbf{P} = \mu\mathbf{R} + (1-\mu)\,\mathbf{r}\mathbf{r}'/n. \tag{3.33}$$

Kageyama (1980) showed that the $\mathbf{A}$ matrix of an efficiency balanced design is

$$\begin{aligned}\mathbf{A} &= \mathbf{R} - \mathbf{N}\mathbf{K}^{-1}\mathbf{N}'\\ &= \mathbf{R}\,(\mathbf{I} - \mathbf{M}_0) - \mathbf{r}\mathbf{r}'/n\end{aligned}$$

and in view of (3.32), we have

$$\mathbf{A} = (1 - \mu)(\mathbf{R} - \mathbf{rr}'/n). \tag{3.34}$$

Remark 3.6: Das and Ghosh (1985) called a design as general efficiency balanced if for some $\theta, s_1, .., s_v \ (> 0)$, **A** is of the form

$$\mathbf{A} = \theta(\mathbf{S} - g^{-1}\mathbf{ss}'), \tag{3.35}$$

where

$$\mathbf{S} = \text{diag}\,(s_1, ..., s_v),\ \mathbf{s} = (s_1, ..., s_v)' \text{ and } \mathbf{s}'\mathbf{1} = g.$$

In particular if $\mathbf{s} = \mathbf{r}$, then the design is efficiency balanced and if $\mathbf{s} = \mathbf{1}$, then the design is variance balanced. The analysis of general efficiency balanced designs is fairly straightforward because a generalized inverse of **A** is

$$\mathbf{A}^- = \theta^{-1}\,\mathbf{S}^{-1}$$

and the adjusted treatment sum of squares is given by $\mathbf{Q}'\mathbf{S}^{-1}\mathbf{Q}/\theta$.

Puri and Nigam (1975a) proved that a design is efficiency balanced if the off diagonal elements of **P** are all proportional to the product of the two relevant replications, i.e.

$$p_{im} = \sum_j n_{ij}\, n_{mj}/k_j \propto r_i r_m \text{ for all } i \neq j = 1, ..., v,$$

with proportionality factor $(1 - \mu)/n$.

Remark 3.7: A design for which the **P**-matrix has all diagonal elements equal and all off-diagonal elements equal is both variance and efficiency balanced (Puri and Nigam, 1975a).

Remark 3.8: A design for which all the off-diagonal elements of **P** are equal but diagonal elements differ is variance balanced but it cannot be efficiency balanced.

Example 3.1: The design (1, 1, 3, 3, 4), (1, 1, 2, 2, 4) and (2, 2, 3, 3, 4) in four treatments is variance balanced but not efficiency balanced.

Example 3.2: The design (1, 1, 3) (2, 3, 5), (1, 4, 5) (1, 2, 6), (1, 2, 4), (3, 4, 6) and (1, 5, 6) in six treatments is an efficiency balanced design but not variance balanced.

We now prove some results on efficiency balanced designs. The following theorem is due to Kageyama (1980). (Proved also by Dey, Singh and Saha, 1981).

Theorem 3.7: In the class of proper, binary designs, BIB design is the only efficiency balanced design if it exists.

Proof : For an efficiency balanced design $D(v, b, \mathbf{r}, k)$, we have

$$\mathbf{P} = \mathbf{NN}'/k = \mu\mathbf{R} + (1 - \mu)\mathbf{rr}'/n,$$

and

$$\mathbf{NN}' = \mu k\mathbf{R} + (1-\mu)k\mathbf{rr}'/n.$$

Comparing the diagonal elements keeping in mind that the *i*-th diagonal element of **NN**′ in case of binary designs is r_i, we have

$$r_i = k\,[\mu r_i + (1-\mu)r_i^2/n]$$

or

$$r_i = (1/k-\mu)n/(1-\mu) \text{ for all } i.$$

Thus a proper, binary, efficiency balanced design is also equi-replicated and by remark 3.7 the design is variance balanced as well. It is well known that the only proper, binary, equi-replicate and variance balanced design is the BIB design, if it exists.

Corollary 3.3: There does not exist an equi-replicated, efficiency balanced, binary design with $b = v + 1$.

Proof: From Theorem 3.7 it follows that a proper, binary, efficiency balanced design is BIB design. Furthermore there does not exist a BIB design with parameters v, $b = v + 1$, r, k and λ. Since the relation $vr = bk$ implies $vr = (v+1)k$, and hence $r - k = k/v < 1$, which leads to a contradiction, because $r - k$ must be an integer. This completes the proof.

As in the case of variance balance, there does not exist a proper and binary, efficiency balanced design with varying replications. (For details the reader may refer to papers by Kageyama (1980, 81, 82)). Kageyama (1980) studied some properties of efficiency balanced designs and derived some bounds on the parameters of these designs. He showed that the inequality $b \geqslant v$ holds also for efficiency balanced designs. This we also prove later in Chapter 7.

Calinski (1971) gave some examples of efficiency balanced designs by trial and error. Puri and Nigam (1975b) gave a systematic procedure of constructing efficiency balanced designs by merging of treatments in BIB and efficiency balanced designs. Their procedure provides a large number of useful efficiency balanced designs. Through their procedure efficiency balanced designs in v^* treatments can be constructed from a given design in v treatments ($v^* < v$) by merging some of the treatments. Designs through merging of treatments are discussed in detail in Chapter 7.

We now take up the case $s = 2$. For $s = 2$, $\mu_1 = 0$ and $\mu_2 = \mu$ (say) and the design is called simple partially efficiency balanced (SPEB) or a *C*-design (Puri and Nigam, 1977a; Saha, 1976). For SPEB-designs ρ_1 contrasts are estimated with full efficiency and ρ_2 contrasts are estimated with efficiency factor μ ($\rho_1 + \rho_2 = v - 1$). The $\mathbf{\Omega}$-matrix of the SPEB designs, using (2.16), is

$$\mathbf{\Omega} = [\mathbf{I} + \mathbf{M}_0/(1-\mu)]\,\mathbf{R}^{-1}. \tag{3.36}$$

We now study some of the known PBIB designs and examine the pattern of their $\mathbf{M}_0$-matrix and show how their analysis gets simplified to a great extent by identifying them as PEB designs.

3.5 PBIB DESIGNS AS PEB DESIGNS

Consider an m-associate PBIB design with parameters v, b, r, k, $\lambda_i(i = 1,\ldots, m)$. If $\mathbf{N}$ denotes the incidence matrix of a PBIB design then it is known from (3.6) that

$$\mathbf{NN}' = r\mathbf{B}_0 + \lambda_1\mathbf{B}_1 + \ldots + \lambda_m\mathbf{B}_m,$$

where $\mathbf{B}_i$'s ($i = 0, 1,\ldots, m$) are the association matrices. Kusumoto (1967) has shown that the concurrence matrix of a PBIB design satisfies the following relation:

$$\mathbf{NN}' = \theta_0\mathbf{L}_0 + \theta_1\mathbf{L}_1 + \ldots + \theta_m\mathbf{L}_m, \tag{3.37}$$

where $\theta_0 = rk$ and θ_i's are the eigen values of $\mathbf{NN}'$ with multiplicities ρ_i and $\mathbf{L}_0 = \mathbf{11}'/v$ and $\mathbf{L}_i$'s are the corresponding idempotent matrices of the algebra generated by association matrices $\mathbf{B}_0, \mathbf{B}_1,\ldots, \mathbf{B}_m$. Hence, $\mathbf{M}$ has the decomposition

$$\mathbf{M} = \sum_{i=0}^{m} \mu_i\mathbf{L}_i,$$

$$\text{where } \mu_i = \theta_i/rk \text{ and } \sum_{i=0}^{m} \mathbf{L}_i = \mathbf{I}.$$

It therefore follows that the m-associate class PBIB design is a PEB design with atmost m efficiency classes having parameters

$$v, b, r, k, \mu_i = \theta_i/rk, \mathbf{L}_i, \rho_i(i = 1,\ldots, m). \tag{3.38}$$

It may be remarked here that the basic contrasts $\mathbf{s}_{ij}$, the eigen vectors of $\mathbf{M}$-matrix, are not necessarily the elementary contrasts and as such may not be of any interest to the experimenter. Our immediate concern, therefore, is to work out the loss of information on elementary contrasts once these are known for basic contrasts.

3.5.1 Efficiency factor of contrasts other than basic contrasts

Consider a PEB design D with m efficiency classes. For such designs we have

$$\mathbf{M}_0 = \sum_i \mu_i \mathbf{L}_i$$

and

$$\begin{aligned}\boldsymbol{\Omega} &= (\mathbf{I} - \mathbf{M}_0)^{-1}\,\mathbf{R}^{-1}\\ &= [\mathbf{I} + \sum_i \mu_i/(1-\mu_i)\,\mathbf{L}_i]\,\mathbf{R}^{-1},\end{aligned}$$

where $\mathbf{L}_i$'s are as given in (3.17).

Thus, having known the losses of information μ_i on basic contrasts of a PEB m-efficiency class design, $\mathbf{\Omega}$ can be readily worked out.

Suppose now the interest is to work out the efficiency factor of a contrast $\boldsymbol{\psi}'\hat{\boldsymbol{\tau}}$ which is not a basic contrast. The efficiency factor for $\boldsymbol{\psi}'\hat{\boldsymbol{\tau}}$ is

$$e(\boldsymbol{\psi}'\hat{\boldsymbol{\tau}}) = \frac{\text{Var}(\boldsymbol{\psi}'\hat{\boldsymbol{\tau}})_0}{\text{Var}(\boldsymbol{\psi}'\hat{\boldsymbol{\tau}})_D} = \frac{\boldsymbol{\psi}'\mathbf{R}^{-1}\boldsymbol{\psi}}{\boldsymbol{\psi}'\mathbf{\Omega}\boldsymbol{\psi}}$$

Once we know $\mathbf{\Omega}$, as described above, $e(\boldsymbol{\psi}'\hat{\boldsymbol{\tau}})$ can be computed easily. We demonstrate this through an example.

Example 3.3: Consider the G.D. design $v = 4$, $b = 4$, $r = 2 = k$, $\lambda_1 = 0$, $\lambda_2 = 1$, $n_1 = 0$, $n_2 = 2$ with block contents (1, 3), (2, 4), (1, 4), (2, 3).

Consider now $\mathbf{s}_1 = (1\ -1\ 0\ 0)'$, $\mathbf{s}_2 = (0\ 0\ 1\ -1)'$ and $\mathbf{s}_3 = (1\ 1\ -1\ -1)'$. These are linearly independent basic contrasts with corresponding losses of information $\mu_1 = \frac{1}{2} = \mu_2$, $\mu_3 = 0$, respectively. We have

$$\mathbf{L}_1 = \tfrac{1}{2}\begin{bmatrix} 1 & -1 & 0 & 0 \\ -1 & 1 & 0 & 0 \\ 0 & 0 & 0 & 0 \\ 0 & 0 & 0 & 0 \end{bmatrix}, \quad \mathbf{L}_2 = \tfrac{1}{2}\begin{bmatrix} 0 & 0 & 0 & 0 \\ 0 & 0 & 0 & 0 \\ 0 & 0 & 1 & -1 \\ 0 & 0 & -1 & 1 \end{bmatrix}$$

and

$$\mathbf{L}_3 = \tfrac{1}{4}\begin{bmatrix} 1 & 1 & -1 & -1 \\ 1 & 1 & -1 & -1 \\ -1 & -1 & 1 & 1 \\ -1 & -1 & 1 & 1 \end{bmatrix}$$

Now

$$(\mathbf{I} - \mathbf{M}_0)^{-1} = \sum_{i=0}^{m} 1/(1 - \mu_i)\,\mathbf{L}_i$$

$$= \tfrac{1}{4}\begin{bmatrix} 5 & -3 & -1 & -1 \\ -3 & 5 & -1 & -1 \\ -1 & -1 & 5 & -3 \\ -1 & -1 & -3 & 5 \end{bmatrix}$$

Consider now $\boldsymbol{\psi} = (1\ 0 - 1\ 0)'$. Then

$$\boldsymbol{\psi}'\boldsymbol{\Omega}\boldsymbol{\psi} = \boldsymbol{\psi}'\,(\mathbf{I} - \mathbf{M}_0)^{-1}\,\mathbf{R}^{-1}\boldsymbol{\psi} = 3/2$$

and $\boldsymbol{\psi}'\mathbf{R}^{-1}\boldsymbol{\psi} = 1$, and $e\,(\boldsymbol{\psi}'\boldsymbol{\tau}) = 2/3$.
Similarly, let $\boldsymbol{\psi} = (1\ 0\ 0 - 1)'$. Then $e\,(\boldsymbol{\psi}'\boldsymbol{\tau}) = 2/3$.
Note that $\boldsymbol{\psi}$ considered above are not basic contrasts.

3.5.2 Analysis of two-associate PBIB designs

The efficiency pattern of some known PBIB designs has been considered by Puri (1975) and Singh (1976). We describe here the analysis of two associate PBIB designs due to Singh (1976).

3.5.2.1 SPEB type PBIB designs

Consider an equi-replicate and proper SPEB design D with parameters $v, b, r, k, \mu, \mathbf{L}$. The relationship $\mathbf{M}_0 = \mu\mathbf{L}$ gives

$$\mathbf{NN}' = rk\,(\mathbf{11}'/v + \mu\mathbf{L}) \tag{3.39}$$

with projection corresponding to zero eigen value being $\mathbf{I} - \mathbf{L} - \mathbf{11}'/v$. For a two-associate PBIB design having one of the eigen values of $\mathbf{NN}'$ zero (throughout we refer to the eigen value other than the simple value rk), the matrix $\mathbf{NN}'$ can be written as

$$\mathbf{NN}' = \theta_0\mathbf{L}_0 + \theta_1\mathbf{L}_1. \tag{3.40}$$

Clearly the two characterizations are same. We thus have the following theorem:

Theorem 3.8: Every two-associate PBIB design $D(v, b, r, k)$ which has only one non-zero eigen value θ_1 of $\mathbf{NN}'$ is SPEB with $\mu = \theta_1/rk$.

A large number of two-associate PBIB designs fall in this category. Linked block and affine resolvable designs, designs obtained through partial geometry (r, k, t), simple lattice, reinforced BIB designs of Das (1958), supplemented efficiency balanced designs of Calinski (1971), semi-regular and singular GD, certain L_i designs with $r + (n - i)\,\lambda_1 - (n - i + 1)\,\lambda_2 = 0$ or $r - i\,\lambda_1 + (i - 1)\,\lambda_2 = 0$, a particular class of triangular designs with $r + (n - 4)\,\lambda_1 - (n - 3)\,\lambda_2 = 0$ or $r - 2\,\lambda_1 + \lambda_2 = 0$, parallel line bio-assay designs of Das and Kulkarni (1966) and Nigam and Boopathy (1985) are all SPEB and can be easily analysed.

3.5.2.2 Other two-associate PBIB designs

For a PBIB design with two associate classes the matrix $\mathbf{NN}'$ is of the form

$$\begin{aligned}\mathbf{NN}' &= \theta_0\mathbf{L}_0 + \theta_1\mathbf{L}_1 + \theta_2\mathbf{L}_2 \\ &= \theta_2\mathbf{I} + (\theta_0 - \theta_2)\,\mathbf{L}_0 + (\theta_1 - \theta_2)\,\mathbf{L}_1,\end{aligned} \tag{3.41}$$

using $\mathbf{L}_2 = \mathbf{I} - \mathbf{L}_0 - \mathbf{L}_1$, and the matrix $\mathbf{M}_0$ is

$$\mathbf{M}_0 = \mu_1 \mathbf{L}_1 + \mu_2 \mathbf{L}_2,$$

where

$$\mu_i = \theta_i / rk,\ i = 1, 2.$$

The $\mathbf{\Omega}$ matrix of two-associate PBIB design can be written as

$$= [\mathbf{I} + \mu_1 \mathbf{L}_1/(1 - \mu_1) + \mu_2 \mathbf{L}_2/(1 - \mu_2)]/r \tag{3.42}$$

$$= rk[(rk - \theta_1)\,\mathbf{I} - \theta_2\,(rk - \theta_1)\,\mathbf{L}_0 + rk\,(\theta_1 - \theta_2)\,\mathbf{L}_1]/ r\,(rk - \theta_1)\,(rk - \theta_2). \tag{3.43}$$

Substituting the value of

$$\mathbf{L}_1 = [\mathbf{NN}' - (\theta_0 - \theta_2)\,\mathbf{L}_0 - \theta_2 \mathbf{I}]/(\theta_1 - \theta_2)$$

and on simplification, keeping in mind that

$$\mathbf{L}_0\,\mathbf{Q} = \mathbf{11}'\mathbf{Q}/v = 0,$$

the sum of squares attributed to treatments becomes

$$\begin{aligned}
\mathbf{Q}'\mathbf{\Omega}\mathbf{Q} &= rk\,[(rk - \theta_1)\,\mathbf{Q}'\mathbf{Q} + \mathbf{Q}'\mathbf{NN}'\mathbf{Q} - \theta_2 \mathbf{Q}'\mathbf{Q}]\,[r(rk - \theta_1)\\
&\quad (rk - \theta_2)]^{-1}\\
&= k[(rk - \theta_1 - \theta_2)\,\mathbf{Q}'\mathbf{Q} + \mathbf{U}'\mathbf{U}]/(rk - \theta_1)\,(rk - \theta_2)\\
&= k[(rk - \theta_1 - \theta_2)\,\sum_{i=1}^{v} Q_i^2 + \sum_{j=1}^{b} u_j^2]/(rk - \theta_1)\,(rk - \theta_2)
\end{aligned} \tag{3.44}$$

where

$$\mathbf{U} = \mathbf{N}'\mathbf{Q} = (u_1, \ldots, u_b)'.$$

As in case of SPEB designs, there is once again no need to obtain the idempotent matrices $\mathbf{L}_i$'s for the analysis of two-associate PBIB designs.

Bose and Shimamoto (1952) classified the two-associate PBIB designs according to their association schemes. We now give the expressions for adjusted treatment sum of squares for the particular classifications.

For GD PBIB designs the eigen values of $\mathbf{NN}'$ other than rk are $\theta_1 = r - \lambda_1$ and $\theta_2 = rk - v\lambda_2$ with respective multiplicities $m\,(n - 1)$ and $m - 1$. The sum of squares due to treatments for GD design is

$$k\,[(\lambda_1 + v\lambda_2 - r)\,\mathbf{Q}'\mathbf{Q} + \mathbf{U}'\mathbf{U}]/v\,(rk - r + \lambda_1)\lambda_2.$$

For triangular type PBIB designs the eigen values of $\mathbf{NN}'$ other than rk are $r + (n - 4)\,\lambda_1 - (n - 3)\,\lambda_2$ and $r - 2\lambda_1 + \lambda_2$ with respective multiplicities $(n - 1)$ and $n(n - 3)/2$. The treatment sum of squares is

$$\frac{k\,[(rk - 2r - (n - 6)\,\lambda_1 + (n - 4)\,\lambda_2)\,\mathbf{Q}'\mathbf{Q} + \mathbf{U}'\mathbf{U}]}{[rk - r - (n - 4)\,\lambda_1 + (n - 3)\,\lambda_2](rk - r + 2\lambda_1 - \lambda_2)}.$$

For latin square type PBIB design the eigen values of $\mathbf{NN}'$ other than rk are $\theta_1 = r + (n - i)\lambda_1 - (n - i + 1)\lambda_2$ and $\theta_2 = r - i\,\lambda_1 + (i - 1)\,\lambda_2$

with respective multiplicities $i(n-1)$ and $(n-i+1)(n-1)$. The treatment sum of squares is

$$\frac{k\,[(rk-2r-(n-2i)\lambda_1+(n-2i+2)\lambda_2\,\mathbf{Q'Q}+\mathbf{U'U}]}{[rk-r-(n-i)\lambda_1+(n-i+1)\lambda_2][rk-r+i\lambda_1-(i-1)\lambda_2]}$$

3.6 PAIRWISE BALANCE

The concept of pairwise balance in block designs is of combinatorial interest and was introduced by Bose and Shrikhande (1960) in the context of constructing mutually orthogonal latin squares. They defined the pairwise balanced designs in the following manner:

Definition 3.5: A binary incomplete block design $D(v, b, \mathbf{r}, \mathbf{k})$ is pairwise balanced if every pair of distinct symbols occurs in λ blocks of the design.

A characterization of pairwise balanced designs is given in the following definition:

Definition 3.6: A block design $\mathbf{N}$ is called combinatorially balanced or pairwise balanced if the off-diagonal elements of $\mathbf{NN'}$ are constant, i.e.,

$$\mathbf{NN'} = \mathbf{D} + \lambda\mathbf{11'}. \tag{3.45}$$

Hedayat and Federer (1974) have shown through counter examples that pairwise balance is neither necessary nor sufficient for a block design to have variance balance. It can be seen easily that if the design is proper then (3.2) reduces to (3.45) and in that case both the concepts of balance coincide.

Pairwise balance is neither necessary nor sufficient for a block design to have efficiency balance as demonstrated in the following theorem.

Theorem 3.9: If a design is pairwise (efficiency) balanced, this does not imply that it is also efficiency (pairwise) balanced.

Proof: By counter examples.

Example 3.4: Consider the following design in six treatments and seven blocks (1, 2, 3, 6), (1, 3, 4, 5), (2, 4, 5, 6), (1, 2, 5), (2, 3, 4), (1, 4, 6) and (3, 5, 6). The design is pairwise balanced as all the off-diagonal elements of $\mathbf{NN'}$ are equal to 2. It can easily be shown that the design is not efficiency balanced.

Example 3.5: Consider the following design in eight treatments and fourteen blocks: (1, 2, 3, 6, 8, 8), (2, 4, 5, 6, 8, 8), (3, 5, 6, 7, 8, 8), (1, 3, 4, 5, 8, 8),

(1, 4, 6, 7, 8, 8), (2, 3, 4, 7, 8, 8), (1, 2, 5, 7, 8, 8), (1, 2, 4), (4, 5, 7), (3, 4, 6), (2, 6, 7), (1, 5, 7), (2, 3, 5) and (1, 3, 7).

It can be shown easily that the design is efficiency balanced with $\mu = 1/7$. It is not pairwise balanced as off-diagonal elements of NN' can be seen to be different.

4. Kronecker Product Designs

4.1 INTRODUCTION

This Chapter is devoted to study a class of designs known as Kronecker Product Designs. These designs are obtained by taking Kronecker product of two or more designs. Vartak (1955, 1959) introduced these designs and proved a general result that if $\mathbf{N}_1$ and $\mathbf{N}_2$ are the incidence matrices of two PBIB designs with m_1 and m_2 associate classes respectively, then $\mathbf{N}=\mathbf{N}_1 \otimes \mathbf{N}_2$, where $\otimes$ denotes the Kronecker product of matrices, is the incidence matrix of a PBIB design with at most $m_1 + m_2 + m_1 m_2$ associate classes. Many of the well known PBIB designs can be shown to belong to this class of designs. Lot of work has been done on this topic since then. The class of Kronecker Product designs is advantageous in several ways. Firstly, the analysis of this class of designs is extremely simple and, as will be seen later, the analysis depends solely upon the analysis of original designs which are simpler in analysis in view of their structural and design properties. Another advantage of Kronecker Product designs is that they have close connection with designs having factorial structure. We study this class of designs in Chapter 5. Before we give a detailed exposition of Kronecker Product designs, some preliminaries are in order.

4.2 SOME PRELIMINARIES

4.2.1 Kronecker product of matrices

Let $\mathbf{A} = (a_{ij})$ and $\mathbf{B} = (b_{kl})$ be respectively $m \times n$ and $p \times q$ matrices, defined over the set of non-negative integers. The Kronecker product $\mathbf{A} \otimes \mathbf{B}$ of matrices $\mathbf{A}$ and $\mathbf{B}$ is defined by

$$\mathbf{A} \otimes \mathbf{B} = \begin{bmatrix} a_{11}\,\mathbf{B} & a_{12}\,\mathbf{B} & \ldots & a_{1n}\,\mathbf{B} \\ a_{21}\,\mathbf{B} & a_{22}\,\mathbf{B} & \ldots & a_{2n}\,\mathbf{B} \\ \vdots & & & \\ a_{m1}\,\mathbf{B} & a_{m2}\,\mathbf{B} & \ldots & a_{mn}\mathbf{B} \end{bmatrix} \tag{4.1}$$

where $a_{ij}\,\mathbf{B}$ $(i = 1, \ldots, m;\ j = 1, \ldots, n)$ is itself a $p \times q$ matrix.

It is clear from the definition that the Kronecker Product of matrices always exists and that $\mathbf{A} \otimes \mathbf{B}$ is an $mp \times nq$ matrix defined over the set of non-negative integers. For some useful results on Kronecker product of matrices reference may be made to Neudecker (1969) and Rao (1973).

4.2.2 The Kronecker product design

If $\mathbf{N}_1$ and $\mathbf{N}_2$ are the incidence matrices of two designs, then the matrix $\mathbf{N} = \mathbf{N}_1 \otimes \mathbf{N}_2$ is the incidence matrix of design D called the Kronecker product design. The design D^* whose incidence matrix is $\mathbf{N}^* = \mathbf{N}_2 \otimes \mathbf{N}_1$ is also a Kronecker product design. The designs D and D^* are structurally the same, i.e. one of them can be obtained from the other simply by renaming the treatments and renumbering the blocks.

Let $\mathbf{N}_1 = (1, 1, \ldots, 1)$ be $1 \times b$ incidence matrix of a design D_1 with a single treatment allocated once to b blocks and $\mathbf{N}_2 = (1, 1, \ldots, 1)'$ be the $v \times 1$ incidence matrix of another design D_2 with v treatments allocated to a single block, then $\mathbf{N} = \mathbf{N}_1 \otimes \mathbf{N}_2 = \mathbf{1}\mathbf{1}'$ is the incidence matrix of order $v \times b$ of the Kronecker product design D. Clearly D is a randomized complete block design in v treatments and b blocks.

In general, if $\mathbf{N}_1$ and $\mathbf{N}_2$ are the incidence matrices of two designs with the concurrence matrices as $\mathbf{N}_1\mathbf{N}_1'$ and $\mathbf{N}_2\mathbf{N}_2'$ then the Kronecker product design has the concurrence matrix $\mathbf{NN}' = (\mathbf{N}_1 \otimes \mathbf{N}_2)(\mathbf{N}_1 \otimes \mathbf{N}_2)'$, which, using the properties of Kronecker product of matrices can be written as

$$\mathbf{NN}' = \mathbf{N}_1\mathbf{N}_1' \otimes \mathbf{N}_2\mathbf{N}_2'. \tag{4.2}$$

We now study some of the important Kronecker product designs. We restrict ourselves to equireplicate and proper designs. Their generalizations are fairly straight forward.

4.3 KRONECKER PRODUCT DESIGNS OBTAINABLE THROUGH TWO BIB DESIGNS

4.3.1 Through Family (A) BIB designs

It was shown by Shrikhande (1962) that family (A) balanced incomplete block designs reproduce themselves. A BIB (v, b, r, k, λ) is said to belong to the family (A) if $b = 4(r - \lambda)$.

Theorem 4 1: If $\mathbf{N}_i$ $(i = 1, 2)$ is the incidence matrix of a family (A) BIB design with parameters v_i, b_i, r_i, k_i, λ_i and $\bar{\mathbf{N}}_i$ is the incidence matrix of the corresponding complementary design, then

$$\mathbf{N} = \mathbf{N}_1 \otimes \mathbf{N}_2 + \bar{\mathbf{N}}_1 \otimes \bar{\mathbf{N}}_2$$

is the incidence matrix of a family (A) BIB design with parameters

$$v = v_1 v_2,\ b = b_1 b_2,\ r = r_1 r_2 + (b_1 - r_1)(b_2 - r_2),$$
$$k = k_1 k_2 + (v_1 - k_1)(v_2 - k_2),\ \lambda = r - b/4. \tag{4.3}$$

Proof: It is clear that $\mathbf{N}$ is a matrix of order $v_1v_2 \times b_1b_2$. Thus $v = v_1v_2$ and $b = b_1b_2$. Noting that $b_i = 4(r_i - \lambda_i)$, and that the parameters of the complementary design are

$$v_i^* = v_i,\ b_i^* = b_i,\ r_i^* = b_i - r_i,\ k_i^* = v_i - k_i,\ \lambda_i^* = b_i - 2r_i + \lambda_i,$$

and $b_i^* = 4(r_i^* - \lambda_i^*)$, $i = 1, 2$, we have

$$\mathbf{N1} = [r_1r_2 + (b_1 - r_1)(b_2 - r_2)]\,\mathbf{1},$$

$$\mathbf{1'N} = [k_1k_2 + (v_1 - k_1)(v_2 - k_2)]\,\mathbf{1'}$$

and

$$\mathbf{NN'} = (b/4)\,\mathbf{I} + (r - b/4)\,\mathbf{1\,1'}.$$

The proof is thus complete.

Example 4.1: Consider the family (A) BIB design $v_1 = b_1 = 4$, $r_1 = k_1 = 3$, $\lambda_1 = 2$. If we choose both $\mathbf{N}_1$ and $\mathbf{N}_2$ as the incidence matrices of this family (A) design, then the incidence matrix $\mathbf{N} = \mathbf{N}_1 \otimes \mathbf{N}_2 + \bar{\mathbf{N}}_1 \otimes \bar{\mathbf{N}}_2$ is the incidence matrix of a family (A) type BIB design with parameters $v = b = 16$, $r = k = 10$, $\lambda = 6$.

4.3.2 Kronecker product of two BIB designs

The following result, due to Vartak (1955), gives the Kronecker product of two BIB designs.

Let $\mathbf{N}_1$ and $\mathbf{N}_2$ be incidence matrices of two BIB designs with parameters $v_1, b_1, r_1, k_1, \lambda_1$ and $v_2, b_2, r_2, k_2, \lambda_2$, respectively. Then $\mathbf{N} = \mathbf{N}_1 \otimes \mathbf{N}_2$ is the incidence matrix of a PBIB design with at most three associate classes having the rectangular association scheme with the parameters

$$\begin{aligned} &v = v_1v_2,\ b = b_1b_2,\ r = r_1r_2,\ k = k_1k_2,\\ &\lambda_{01} = r_1\lambda_2,\ \lambda_{10} = r_2\lambda_1,\ \lambda_{11} = \lambda_1\lambda_2, \qquad (4.4)\\ &n_{01} = v_2 - 1,\ n_{10} = v_1 - 1,\ n_{11} = (v_1 - 1)(v_2 - 1). \end{aligned}$$

The p^i_{jk} parameters given by $\mathbf{P}_i$ matrices, where $i, j, k = 1, 2, 3$, are

$$\mathbf{P}_1 = (p^1_{jk}) = \begin{bmatrix} v_2 - 2 & 0 & 0 \\ 0 & 0 & v_1 - 1 \\ 0 & v_1 - 1 & (v_1 - 1)(v_2 - 2) \end{bmatrix},$$

$$\mathbf{P}_2 = (p^2_{jk}) = \begin{bmatrix} 0 & 0 & v_2 - 1 \\ 0 & v_1 - 2 & 0 \\ v_2 - 1 & 0 & (v_1 - 2)(v_2 - 1) \end{bmatrix},$$

$$\mathbf{P}_3 = (p^3_{jk}) = \begin{bmatrix} 0 & 1 & v_2-2 \\ 1 & 0 & v_1-2 \\ v_2-2 & v_1-2 & (v_1-2)\ (v_2-2) \end{bmatrix}.$$

The rectangular association scheme was introduced by Vartak (1955). We arrange the $v = v_1 v_2$ treatments in the form of a rectangular array with v_1 rows and v_2 columns. Two treatments are first associates if they appear in the same row of the array, second associates if they appear in the same column, otherwise they are third associates.

It may be recalled that the concurrence matrix of a BIB design with parameters (v, b, r, k, λ) can be written as

$$\begin{aligned} \mathbf{NN}' &= (r-\lambda)\,\mathbf{I} + \lambda\mathbf{11}' \\ &= (rk)\left(\frac{1}{v}\,\mathbf{11}'\right) + (r-\lambda)\left[\mathbf{I} - \left(\frac{1}{v}\right)\mathbf{11}'\right] \\ &= \theta_0\mathbf{L}_0 + \theta_1\mathbf{L}_1 \end{aligned}$$

where

$$\theta_0 = rk,\ \theta_1 = r - \lambda,$$

and

$$\mathbf{L}_0 = \left(\frac{1}{v}\right)\mathbf{11}' \text{ and } \mathbf{L}_1 = \mathbf{I} - \left(\frac{1}{v}\right)\mathbf{11}'.$$

We prove the following theorem which is useful as it gives the eigen values and idempotent matrices of the rectangular design in a fairly simple and straight forward manner.

Theorem 4.2: The eigen values of $\mathbf{NN}'$ of a rectangular PBIB design with parameters (4.4) are

$$\theta_{x_1x_2} = \theta_{x_1}\,\theta_{x_2} \text{ with multiplicities}$$

$$\rho_{x_1x_2} = \rho_{x_1}\,\rho_{x_2}$$

and

$$\mathbf{L}_{x_1x_2} = \mathbf{L}_{x_1} \otimes \mathbf{L}^*_{x_2}$$

$$x_1,\ x_2 = 0,\ 1$$

where θ_{x_1} and $\theta^*_{x_2}$ are the eigen values of $\mathbf{N}_1\mathbf{N}_1'$ and $\mathbf{N}_2\mathbf{N}_2'$ with multiplicities ρ_{x_1} and $\rho^*_{x_2}$, respectively and $\mathbf{L}_{x_1}$ and $\mathbf{L}^*_{x_2}$ are the corresponding idempotent matrices.

Proof: We note that

$$\mathbf{NN}' = \mathbf{N}_1\mathbf{N}_1' \otimes \mathbf{N}_2\mathbf{N}_2'$$

$$= (\theta_0 \mathbf{L}_0 + \theta_1 \mathbf{L}_1) \otimes (\theta_0^* \mathbf{L}_0^* + \theta_1^* \mathbf{L}_1^*)$$

$$= \theta_0\theta_0^* \mathbf{L}_0 \otimes \mathbf{L}_0^* + \theta_1\theta_0^* \mathbf{L}_1 \otimes \mathbf{L}_0^* + \theta_0\theta_1^* \mathbf{L}_0 \otimes \mathbf{L}_1^* + \theta_1\theta_1^* \mathbf{L}_1 \otimes \mathbf{L}_1^* .$$

Then the matrix $\mathbf{NN}'$ can be written as

$$\mathbf{NN}' = \Sigma\theta_{x_1x_2} \mathbf{L}_{x_1x_2} .$$

Using $r_l k_l = r_l + (v_l - 1)\lambda_l$, $l = 1, 2$, and (4.4) we have

$$\theta_{00} = r_1k_1r_2k_2 = rk,$$

$$\theta_{01} = r_1k_1(r_2 - \lambda_2)$$

$$= [r_1 + (v_1 - 1)\lambda_1](r_2 - \lambda_2)$$

$$= r - \lambda_{01} + (v_1 - 1)(\lambda_{10} - \lambda_{11}),$$

$$\theta_{10} = r_2k_2(r_1 - \lambda_1) = r - \lambda_{10} + (v_2 - 1)(\lambda_{01} - \lambda_{11}),$$

and

$$\theta_{11} = (r_1 - \lambda_1)(r_2 - \lambda_2) = r - \lambda_{10} - \lambda_{01} + \lambda_{11}.$$

Also,

$$\mathbf{L}_{00} = (\mathbf{11}'/v_1) \otimes (\mathbf{11}'/v_2) = \mathbf{11}'/v$$

$$\mathbf{L}_{01} = (\mathbf{11}'/v_1) \otimes (\mathbf{I} - \mathbf{11}'/v_2)$$

$$\mathbf{L}_{10} = (\mathbf{I} - \mathbf{11}'/v_1) \otimes (\mathbf{11}'/v_2)$$

$$\mathbf{L}_{11} = (\mathbf{I} - \mathbf{11}'/v_1) \otimes (\mathbf{I} - \mathbf{11}'/v_2)$$

are the corresponding idempotent matrices.

The multiplicities of various eigen values are the traces of corresponding idempotent matrices and are given by

$$\rho_{00} = 1,\ \rho_{01} = (v_2 - 1),\ \rho_{10} = (v_1 - 1),\ \rho_{11} = (v_1 - 1)(v_2 - 1)$$

This completes the proof.

Example 4.2: Let $\mathbf{N}_1$ be the incidence matrix of a BIB design with parameters

$$v_1 = b_1 = 4,\ r_1 = k_1 = 3,\ \lambda_1 = 2$$

and $\mathbf{N}_2$ be the incidence matrix of another BIB design with parameters

$$v_2 = b_2 = 3,\ r_2 = k_2 = 2,\ \lambda_2 = 1.$$

Then

$$\mathbf{N} = \mathbf{N}_1 \otimes \mathbf{N}_2$$

is the incidence matrix of a rectangular PBIB design with the parameters

$$v = b = 12,\ r = k = 6,\ \lambda_{01} = 3,\ \lambda_{10} = 4,\ \lambda_{11} = 2,$$

$$n_{01} = 2,\ n_{10} = 3,\ n_{11} = 6.$$

The $\mathbf{\Omega}$ matrix for a rectangular PBIB design can be written as

$$\mathbf{\Omega} = [\mathbf{I} + \sum_{d=1}^{2} \{ \sum_{x_1+x_2=d} \mu_{x_1x_2} \mathbf{L}_{x_1x_2} /(1 - \mu_{x_1x_2})\}]/r,$$

where as usual $\mu_{x_1x_2} = \theta_{x_1x_2}/rk$.
Using

$$\sum_{d=0}^{2} [\sum_{x_1+x_2=d} \mathbf{L}_{x_1x_2}] = \mathbf{I}_v$$

and on simplification we have

$$\mathbf{\Omega} = \mathbf{11}'/rv + \sum_{d=1}^{2} [\sum_{x_1+x_2=d} \mathbf{L}_{x_1x_2}/r\,(1 - \mu_{x_1x_2})]$$

Let

$$\mathbf{Q} = [Q_{0,0}\; Q_{0,1} \ldots Q_{0,v_2-1}\; Q_{1,0}\; Q_{1,1} \ldots Q_{1,v_2-1} \ldots Q_{v_1-1,0}$$
$$Q_{v_1-1,1} \ldots Q_{v_1-1,v_2-1}],$$

be the vector of adjusted treatment totals, where

$$Q_{ij}\; (i = 0, 1, \ldots, v_1 - 1 ; j = 0, 1, \ldots, v_2 - 1)$$

denotes the adjusted yield of a treatment in i-th row and j-th column. Then the adjusted treatment sum of squares is

$$\mathbf{Q}'\mathbf{\Omega}\mathbf{Q} = \sum_{d=1}^{2} [\sum_{x_1+x_2=d} \mathbf{Q}' \mathbf{L}_{x_1x_2} \mathbf{Q}]/r\,(1 - \mu_{x_1x_2})$$

$$= \Sigma\, Q_{0j}^2/rv_1\,(1 - \mu_{01}) + \Sigma\, Q_{i0}^2/rv_2\,(1 - \mu_{10})$$
$$+ [\sum_{i,j} Q_{ij}^2 - \sum_{i} Q_{i0}^2/v_2 - \Sigma\, Q_{0j}^2/v_1]/r\,(1 - \mu_{11})$$

4.4 KRONECKER PRODUCT OF TWO PBIB DESIGNS

Let $\mathbf{N}_1$ and $\mathbf{N}_2$ be the incidence matrices of two PBIB designs with m_1 and m_2 associate classes. Vartak (1955) showed that

$$\mathbf{N} = \mathbf{N}_1 \otimes \mathbf{N}_2$$

is the incidence matrix of a PBIB design with atmost $m_1 + m_2 + m_1m_2$ associate classes. Surendran (1968) established Vartak's result more elegantly with the help of association matrices.

It may be recalled that the concurrence matrix of an m-associate PBIB design with parameters $v, b, r, k, \lambda_i, n_i, i = 1, \ldots, m$, can be written as

$$\mathbf{NN}' = \sum_{i=0}^{m} \theta_i\, \mathbf{L}_i$$

where $\theta_0 = rk$, θ_i, $i = 1, ..., m$ are the eigen values of $\mathbf{NN}'$ and $\mathbf{L}_0 = \frac{1}{v}\mathbf{11}'$ and $\mathbf{L}_i$ are the corresponding idempotent matrices with ranks ρ_i. We prove the following theorem:

Theorem 4.3: Let $\mathbf{N}_1$ and $\mathbf{N}_2$ be the incidence matrices of two PBIB designs with parameters

$$v, b, r, k, \lambda_i, n_i, \theta_i, \rho_i, \mathbf{L}_i, i = 1,..., m_1$$

and

$$v^*, b^*, r^*, k^*, \lambda_j^*, n_j^*, \theta_j^*, \rho_j^*, \mathbf{L}_j^*, j = 1,..., m_2.$$

Then $\mathbf{N} = \mathbf{N}_1 \otimes \mathbf{N}_2$ is the incidence matrix of a PBIB design whose concurrence matrix $\mathbf{NN}'$ has the eigen values and idempotent matrices as

$$\theta_{ij} = \theta_i \theta_j^*$$

with multiplicities $\rho_{ij} = \rho_i \rho_j^*$ (4.5)
and

$$\mathbf{L}_{ij} = \mathbf{L}_i \otimes \mathbf{L}_j^*, \ i = 0, 1,..., m_1; j = 0, 1,..., m_2 \text{ respectively.}$$

Proof: Note that

$$\mathbf{NN}' = \mathbf{N}_1\mathbf{N}_1' \otimes \mathbf{N}_2\mathbf{N}_2'$$

$$= [rk\,(\mathbf{11}'/v) + \sum_{i=1}^{m_1} \theta_i \mathbf{L}_i] \otimes [r^*k^*(\mathbf{11}'/v^*) + \sum_{j=1}^{m_2} \theta_j^* \mathbf{L}_j^*]$$

$$= rr^*kk^*\,(\mathbf{11}' \otimes \mathbf{11}')/vv^* + \sum_{i=1}^{m_1} r^*k^*\theta_i \mathbf{L}_i \otimes (\mathbf{11}'/v^*)$$

$$+ \sum_{j=1}^{m_2} rk\theta_j^*\,(\mathbf{11}'/v) \otimes \mathbf{L}_j^* + \sum_{i=1}^{m_1} \sum_{j=1}^{m_2} \theta_i \theta_j^* \mathbf{L}_i \otimes \mathbf{L}_j^*$$

$$= rr^*kk^*\,(\mathbf{11}'/vv^*) + \sum_{i=1}^{m_1} \theta_{i0}\mathbf{L}_{i0} + \sum_{j=1}^{m_2} \theta_{0j}\mathbf{L}_{0j} + \sum_{i=1}^{m_1} \sum_{j=1}^{m_2} \theta_{ij}\mathbf{L}_{ij}.$$

The multiplicities $\rho_{ij} = \rho_i \rho_j^*$ are easily established. The proof is thus complete.

Example 4.3: Let $\mathbf{N}_1$ be the incidence matrix of a BIB design with parameters $v = b = 3$, $r = k = 2$, $\lambda = 1$ and $\mathbf{N}_2$ be the incidence matrix of a group divisible design with parameters

$$v^* = 6, b^* = 4, r^* = 2, k^* = 3, \lambda_1^* = 0, \lambda_2^* = 1, n_1^* = 1, n_2^* = 4.$$

The eigen values of $\mathbf{N}_1\mathbf{N}_1'$ are $\theta_0 = rk = 4$, $\theta_1 = (r - \lambda) = 1$ with multiplicities $\rho_0 = 1$, $\rho_1 = 2$ respectively, and the corresponding idempotent matrices are

$\mathbf{L}_0 = (1/3)\,\mathbf{11}'$, $\mathbf{L}_1 = (\mathbf{I} - \mathbf{11}'/3)$ and the eigen values of $\mathbf{N}_2\mathbf{N}_2'$ are

$$\theta_0^* = r^*k^* = 6,\ \theta_1^* = r^* - \lambda_1^* = 2$$

and

$$\theta_2^* = r^*k^* - v^*\lambda_2^* = 0$$

with multiplicities respectively,

$$\rho_0^* = 1,\ \rho_1^* = 3,\ \rho_2^* = 2,$$

and

$$\mathbf{L}_0^* = \mathbf{11}'/6,\ \mathbf{L}_1^* = \mathbf{11}'/3 \otimes (\mathbf{I} - \mathbf{11}'/2)$$
$$\mathbf{L}_2^* = (\mathbf{I} - \mathbf{11}'/3) \otimes \mathbf{11}'/2.$$

The eigen values of $\mathbf{NN}'$ are $\theta_{00} = 24$, $\theta_{01} = 8$, $\theta_{02} = 0$, $\theta_{10} = 6$, $\theta_{11} = 2$, $\theta_{12} = 0$, with respective multiplicities 1, 3, 2, 2, 6 and 4. The corresponding idempotent matrices are

$$\mathbf{11}'/18,\ (\mathbf{11}'/3) \otimes (\mathbf{11}'/3) \otimes (\mathbf{I} - \mathbf{11}'/2),$$
$$(\mathbf{11}'/3) \otimes (\mathbf{I} - \mathbf{11}'/3) \otimes (\mathbf{11}'/2),$$
$$(\mathbf{I} - \mathbf{11}'/3) \otimes \mathbf{11}'/6,$$
$$(\mathbf{I} - \mathbf{11}'/3) \otimes (\mathbf{11}'/3) \otimes (\mathbf{I} - \mathbf{11}'/2),$$

and

$$(\mathbf{I} - \mathbf{11}'/3) \otimes (\mathbf{I} - \mathbf{11}'/3) \otimes (\mathbf{11}'/2).$$

Clearly for this design $\mathbf{NN}'$ has only four distinct eigen values as

$$\theta_{12} = \theta_{02} = 0.$$

The design has infact reduced association scheme and is the generalized right angular type. This class of designs with reduced association classes is discussed in detail in Section 4.6.

4.5 KRONECKER PRODUCT OF *h* BIB DESIGNS

Hinkelmann and Kempthorne (1963) and Hinkelmann (1964) extended the results of Vartak (1955) by taking the Kronecker product of h BIB designs D_j $(v_j, b_j, r_j, k_j, \lambda_j)$. They proved that the Kronecker product design is PBIB with $2^h - 1$ associate classes and termed the association scheme as extended group divisible (EGD) which may be defined as given below:

Definition 4.1: Let $v = \prod_{j=1}^{h} v_j$ treatments be designated by h-tuple $(i_1, \ldots, i_h)$,

$i_j = 0, 1, \ldots, v_j - 1; j = 1, \ldots, h$. Any two treatments $(i_1, \ldots, i_h)$ and $(i_1', \ldots, i_h')$ are $(X_1, \ldots, X_h)$-th associate if $i_j = i_j'$ for $j = 1, \ldots, h - (X_1, \ldots, X_h)$, where $X_j = 1$ if $i_j \neq i_j'$ and $X_j = 0$ otherwise. Let $\lambda_{X_1 \ldots X_h}$ be the number of blocks in which every $(X_1, \ldots, X_h)$-th associate pair of treatments occurs together. The parameters of an EGD/$2^h - 1$ PBIB design can be written as

$$v = \prod_{j=1}^{h} v_j, \quad b = \prod_{j=1}^{h} b_j, \quad r = \prod_{j=1}^{h} r_j, \quad k = \prod_{j=1}^{h} k_j,$$

$$\lambda_{X_{1\bullet}} \quad , \quad n_{X_1 \ldots X_h}.$$

This scheme was termed factorial (F_h-type) association scheme by Yamamoto, Fuji and Hamada (1965) and binary number association scheme by Paik and Federer (1973).

For $h = 3$ the treatments can be indexed (i_1, i_2, i_3), $i_l = 0, 1, \ldots, v_l - 1$, $l = 1, 2, 3$, and we get EGD/$2^3 - 1$ PBIB design with parameters $v = v_1 v_2 v_3$, $b, r, k, \lambda_{X_1 X_2 X_3}, n_{X_1 X_2 X_3}$, where

$$\lambda_{100} = \lambda_1 r_2 r_3 \qquad n_{100} = v_1 - 1$$

$$\lambda_{110} = \lambda_1 \lambda_2 r_3 \qquad n_{110} = (v_1 - 1)(v_2 - 1)$$

$$\lambda_{111} = \lambda_1 \lambda_2 \lambda_3 \qquad n_{111} = (v_1 - 1)(v_2 - 1)(v_3 - 1)$$

with analogous definitions for $\lambda_{010}, \lambda_{001}, \lambda_{101}, \lambda_{011}, n_{010}, n_{001}, n_{101}$ and n_{011}.

The eigen values of $\mathbf{NN}'$ can be worked out as

$$\theta_{000} = rk$$

$$\begin{aligned}\theta_{100} &= (r_1 - \lambda_1) r_2 k_2 r_3 k_3 \\ &= (r_1 - \lambda_1)[r_2 + (v_2 - 1)\lambda_2][r_3 + (v_3 - 1)\lambda_3] \\ &= r - \lambda_{100} + (v_2 - 1)(\lambda_{010} - \lambda_{110}) + (v_3 - 1)(\lambda_{001} - \lambda_{101}) \\ &\qquad + (v_2 - 1)(v_3 - 1)(\lambda_{011} - \lambda_{111})\end{aligned}$$

$$\begin{aligned}\theta_{110} &= (r_1 - \lambda_1)(r_2 - \lambda_2) r_3 k_3 \\ &= (r_1 - \lambda_1)(r_2 - \lambda_2)[r_3 - \lambda_3(v_3 - 1)] \\ &= r - \lambda_{100} - \lambda_{010} + \lambda_{110} - (v_3 - 1)(\lambda_{001} - \lambda_{101} - \lambda_{011} + \lambda_{111})\end{aligned}$$

and
$$\begin{aligned}\theta_{111} &= (r_1 - \lambda_1)(r_2 - \lambda_2)(r_3 - \lambda_3) \\ &= r - \lambda_{001} - \lambda_{010} - \lambda_{100} + \lambda_{011} + \lambda_{101} + \lambda_{110} - \lambda_{111}\end{aligned}$$

with multiplicities 1, $(v_1 - 1)$, $(v_1 - 1)(v_2 - 1)$ and $(v_1 - 1)(v_2 - 1)(v_3 - 1)$, respectively. Similar expressions can be written for other θ's.

The corresponding idempotent matrices are

$$\mathbf{L}_{000} = (\mathbf{11}'/v_1) \otimes (\mathbf{11}'/v_2) \otimes (\mathbf{11}'/v_3)$$

$$\mathbf{L}_{100} = (\mathbf{I} - \mathbf{11}'/v_1) \otimes (\mathbf{11}'/v_2) \otimes (\mathbf{11}'/v_3)$$

$$\mathbf{L}_{110} = (\mathbf{I} - \mathbf{11}'/v_1) \otimes (\mathbf{I} - \mathbf{11}'/v_2) \otimes (\mathbf{11}'/v_3)$$
$$\mathbf{L}_{111} = (\mathbf{I} - \mathbf{11}'/v_1) \otimes (\mathbf{I} - \mathbf{11}'/v_2) \otimes (\mathbf{I} - \mathbf{11}'/v_3)$$

Similar expressions can be written for other $\mathbf{L}_i$'s. The $\mathbf{\Omega}$ matrix for EGD/$2^3 - 1$ can be written as

$$\mathbf{\Omega} = \mathbf{11}'/rv + \sum_{d=1}^{3} [\sum_{X_1+X_2+X_3=d} \mathbf{L}_{X\,X_2X_3}/r\,(1 - \mu_{X_1X_2X_3})]$$

If $\mathbf{Q}$ denotes the vector of adjusted treatment totals, then the sum of squares attributed to treatments can be written as

$$\mathbf{Q}'\,\mathbf{\Omega}\,\mathbf{Q} = \sum_{d=1}^{3} [\sum_{X_1+X_2+X_3=d} \mathbf{Q}'\,\mathbf{L}_{X_1X_2X_3}\,\mathbf{Q}/r\,(1 - \mu_{X_1X_2X_3})]$$

The quantities $\mathbf{Q}'\mathbf{L}_{X_1X_2X_3}\,\mathbf{Q}$ can be simplified to a great extent by substituting the values of $\mathbf{L}_{X_1X_2X_3}$. Let Q_{ijk} $(i = 0, 1,..., v_1 - 1;\ j = 0, 1,..., v_2 - 1;\ k = 0, 1,..., v_3 - 1)$ denote the adjusted yield of (i, j, k)-th treatment, and

$$\mathbf{Q}_{i00} = \sum_{j,k} Q_{ijk};\quad Q_{ij0} = \sum_{k} Q_{ijk} \text{ etc.}$$

Then

$$\mathbf{Q}'\,\mathbf{L}_{100}\,\mathbf{Q} = \sum_{i} Q_{i..}^2/v_2v_3$$

$$\mathbf{Q}'\,\mathbf{L}_{110}\,\mathbf{Q} = [\sum_{ij} Q_{ij.}^2 - \sum_{i} Q_{i..}^2/v_2 - \sum_{j} Q_{.j.}^2/v_1]/v_3$$

$$\mathbf{Q}'\,\mathbf{L}_{111}\,\mathbf{Q} = \sum_{i,j,k} Q_{ijk}^2 - \sum_{i,j} Q_{ij.}^2/v_3 - \sum_{i,k} Q_{i.k}^2/v_2 -$$

$$\sum_{j,k} Q_{.jk}^2/v_1 + \sum_{i} Q_{i..}^2/v_2v_3 + \sum_{j} Q_{.j.}^2/v_1v_3 + \sum_{k} Q_{..k}^2/v_1v_2$$

and so on.

4.6 KRONECKER PRODUCT DESIGNS WITH REDUCED ASSOCIATION SCHEMES

We now study Kronecker product PBIB designs with t_1 associate classes obtainable from Kronecker product designs with $t \geqslant t_1$ associate classes. Results on reduction of associate classes have been proved by various workers in terms of PBIB design parameters. Vartak (1955) showed that a rectangular PBIB design, obtained as a Kronecker product of two BIB designs, with three associate classes reduces to a group divisible PBIB with two associate classes if $\lambda_{01} = \lambda_{11}$ or $\lambda_{10} = \lambda_{11}$ and to $\mathbf{L}_2$ type two associate class PBIB design if $\lambda_{01} = \lambda_{10}$ and $v_1 = v_2$. The right angular and generalized right angular schemes with four associate classes were shown by Kusumoto (1976) to be reducible from the Kronecker product of group divisible and BIB designs. Kageyama (1972) established that the necessary and sufficient condition for the Kronecker product of three BIB designs, which has at most seven associate classes, to be reducible to cubic association scheme of Raghavarao and Chandrasekhara Rao (1964) with three associate classes is $v_l = v$ and $k_l = k$ for all $l = 1, 2, 3$.

We now give some of these results and obtain the eigen values and idempotent matrices of the concurrence matrix of the resulting design.

4.6.1 Group divisible PBIB designs

Rectangular association scheme is always reducible to group divisible scheme if we take $\lambda_{10} = \lambda_{11}$ and combine the second associates and third associates of rectangular association scheme, i.e. we call two treatments of the same row (called group) to be first associates and the remaining to be second associates.

By substituting $\lambda_{01} = \lambda_1'$ and $\lambda_{10} = \lambda_{11} = \lambda_2'$ we get the parameters of the group divisible design as

$$v = v_1 v_2,\ b = b_1 b_2,\ r = r_1 r_2,\ k = k_1 k_2,\ \lambda_1' = \lambda_{01},\ \lambda_2' = \lambda_{10} = \lambda_{11}$$

and

$$n_1 = v_2 - 1,\ n_2 = (v_1 - 1) + (v_1 - 1)(v_2 - 1) = (v_1 - 1)\, v_2. \tag{4.6}$$

The eigen values of $\mathbf{NN}'$ can be obtained easily by substituting $\lambda_{01} = \lambda_1'$ and $\lambda_{10} = \lambda_{11} = \lambda_2'$ in Theorem 4.2. These are $\theta_0 = \theta_{00} = rk$, $\theta_1 = \theta_{01} = \theta_{11} = r - \lambda_1'$, $\theta_2 = \theta_{10} = r - \lambda_2 + (v_2 - 1)(\lambda_1 - \lambda_2) = rk - v\lambda_2'$ with multiplicities $\rho_0 = 1$, $\rho_1 = (v_2 - 1) + (v_1 - 1)(v_2 - 1) = v_1(v_2 - 1)$ and $\rho_2 = (v_1 - 1)$. The corresponding idempotent matrices are $\mathbf{L}_0 = \mathbf{L}_{00} = \mathbf{11}'/v$, $\mathbf{L}_1 = \mathbf{L}_{01} + \mathbf{L}_{11}$ and $\mathbf{L}_2 = \mathbf{L}_{10}$ and we have the following theorem:

Theorem 4.4: The eigen values of $\mathbf{NN}'$ of the group divisible PBIB design with parameters (4.6) are

$$\theta_0 = rk,\ \theta_1 = r - \lambda_1' \text{ and } \theta_2 = rk - v\lambda_2'$$

with multiplicities $\rho_0 = 1$, $\rho_1 = v_1(v_2 - 1)$ and $\rho_2 = (v_1 - 1)$. The corresponding idempotent matrices are

$$\mathbf{L}_0 = \mathbf{11}'/v,\ \mathbf{L}_1 = \mathbf{I} \otimes (\mathbf{I} - \mathbf{11}'/v_2)$$

and

$$\mathbf{L}_2 = (\mathbf{I} - \mathbf{11}'/v_1) \otimes \mathbf{11}'/v_2.$$

Similar results can be obtained by putting

$$\lambda_{01} = \lambda_{11} = \lambda_1' \text{ and } \lambda_{10} = \lambda_2'.$$

4.6.2 $\mathbf{L}_2$ – PBIB designs

If $\lambda_{01} = \lambda_{10}$ and $v_1 = v_2 = m$, then rectangular association scheme reduces to L_2 scheme. The parameters of L_2 PBIB design are

$$v = m^2,\ b = b_1 b_2,\ r = r_1 r_2,\ k = k_1 k_2,\ \lambda_1',\ \lambda_2',\ n_1 = 2(m-1),\ n_2 = (m-1)^2. \tag{4.7}$$

The following result follows from Theorem 4.2 by putting $v_1 = v_2 = m$, $\lambda_{10} = \lambda_{01} = \lambda_1'$, $\lambda_{11} = \lambda_2'$ and combining first and second associate classes.

Theorem 4.5: For a L_2 PBIB design with parameters (4.7), the eigen values of $\mathbf{NN}'$ are

$$\theta_0 = rk,\ \theta_1 = r + (m-2)\lambda_1' - (m-1)\lambda_2'$$

and

$$\theta_2 = r - 2\lambda_1' + \lambda_2'$$

with respective multiplicities 1, $2(m-1)$, $(m-1)^2$ and the corresponding idempotent matrices are

$$\mathbf{L}_0 = \mathbf{11}'/v,\ \mathbf{L}_1 = (\mathbf{I} - \mathbf{11}'/m) \otimes (\mathbf{11}'/m)$$
$$+ (\mathbf{11}'/m) \otimes (\mathbf{I} - \mathbf{11}'/m)$$

and

$$\mathbf{L}_2 = (\mathbf{I} - \mathbf{11}'/m) \otimes (\mathbf{I} - \mathbf{11}'/m).$$

4.6.3 Cubic PBIB designs

It was shown by Kageyama (1972) that the EGD/$2^3 - 1$ PBIB design obtained through the Kronecker products of three BIB designs is reducible to three associate cubic PBIB design of Raghavarao and Chandrasekhara rao (1964) if $v_j = v$ and $k_j = k$ for all $j = 1, 2, 3$.

Since $v_j = v$ and $k_j = k$, we get for the j-th BIB design $\lambda_j(v-1) = r_j(k-1)$. Using this, we obtain

$$\lambda_j r_{j'} = \lambda_{j'} r_j;\ j, j' = 1, 2, 3.$$

This gives

$$\lambda_{100} = \lambda_{010} = \lambda_{001} = \lambda_1'$$
$$\lambda_{110} = \lambda_{101} = \lambda_{011} = \lambda_2'$$
$$\lambda_{111} = \lambda_3'.$$

The parameters of the cubic design can be written as v^3, $b = b_1 b_2 b_3$, $r = r_1 r_2 r_3$, k^3, λ_1', λ_2', λ_3', $n_1 = 3(v-1)$, $n_2 = 3(v-1)^2$ and $n_3 = (v-1)^3$. The eigen values of $\mathbf{NN}'$ other than rk are

$$\theta_1 = \theta_{100} = \theta_{010} = \theta_{001} = (r - \lambda_1') + 2(v-1)(\lambda_1' - \lambda_2')$$
$$+ (v-1)^2(\lambda_2' - \lambda_3')$$
$$\theta_2 = \theta_{110} = \theta_{101} = \theta_{011} = r - 2\lambda_1' + \lambda_2' + (v-1)(\lambda_1' - 2\lambda_2' + \lambda_3')$$
$$\theta_3 = \theta_{111} = r - 3(\lambda_1' - \lambda_2') - \lambda_3'$$

with multiplicities $\rho_1 = 3(v-1)$, $\rho_2 = 3(v-1)^2$ and $\rho_3 = (v-1)^3$.

The corresponding idempotent matrices are

$$\mathbf{L}_i = \sum_{X_1+X_2+X_3=i} \mathbf{L}_1^{X_1} \otimes \mathbf{L}_2^{X_2} \otimes \mathbf{L}_3^{X_3},\ X_i = 0, 1;\ i = 1, 2, 3.$$

4.6.4 Hyper-cubic PBIB designs

Hyper-cubic PBIB designs were obtained by Kusumoto (1965) as a generalization of cubic design by taking in an EGD $(2^h - 1)$, $v_j = v$ and $k_j = k$ for all $j = 1, \ldots, h$. Here v^h treatments are denoted by $(i_1, \ldots, i_h)$. Any two treatments $(i_1, \ldots, i_h)$ and $(i_1', \ldots, i_h')$ are j-th associates if the number of non-zero differences $i_j \quad i_j'$ is j. The parameters of the design are v^h, $b = \Pi b_j$, $r = \Pi r_j$, k^h, λ_j', $n_j = \binom{h}{j}(v-1)^j$; $j = 1, \ldots, h$. The eigen values and the idempotent matrices of the spectral decomposition of $\mathbf{NN}'$ can easily be obtained from EGD/$2^h - 1$.

4.6.5 Generalized right angular and right angular PBIB designs

The four associate-class right angular and generalized right angular schemes were introduced by Tharthare (1963, 65). The $v_1 v_2 v_3$ treatments are denoted by ordered triplet (i_1, i_2, i_3), $i_j = 0, 1, \ldots, v_j - 1$; $j = 1, 2, 3$. For the treatment (i_1, i_2, i_3), first associates are those that differ in the third position, second associates are those that differ in the second position while being same or different in the third position, third associates are those that have the same second position, a different first position and the same or different third position and the others are fourth associates.

Kageyama (1974) showed that EGD/$2^3 - 1$ is reducible to generalized right angular if

$$\lambda_1' = \lambda_{010},\ \lambda_2' = \lambda_{110},\ \lambda_3' = \lambda_{011},\ \lambda_4' = \lambda_{111} = \lambda_{101}.$$

The parameters of the reduced design are

$$v_1 v_2 v_3,\ b = b_1 b_2 b_3,\ r = r_1 r_2 r_3,\ k = k_1 k_2 k_3,\ \lambda_1', \lambda_2', \lambda_3', \lambda_4',\ n_1 = n_{010} = v_2 - 1,$$
$$n_2 = n_{100} + n_{110} = (v_1 - 1)\, v_2,\ n_3 = n_{001} + n_{011} = v_2 (v_3 - 1),$$
$$n_4 = n_{101} + n_{111} = (v_1 - 1)(v_3 - 1)\, v_2.$$

The eigen values of $\mathbf{NN}'$ are

$$\theta_0 = \theta_{000} = rk$$
$$\theta_1 = \theta_{100} = r - \lambda_1' + v_2 (v_3 - 1)(\lambda_3' - \lambda_4') + v_2 (\lambda_1' - \lambda_2')$$
$$\theta_2 = \theta_{010} = \theta_{110} = \theta_{011} = \theta_{111} = r - \lambda_1'$$
$$\theta_3 = \theta_{001} = r - \lambda_1' + v_2 [\lambda_1' - \lambda_3' + (v_1 - 1)(\lambda_2' - \lambda_4')]$$

and

$$\theta_4 = \theta_{101} = r - \lambda_1' + v_2 (\lambda_1' - \lambda_2' - \lambda_3' + \lambda_4').$$

The multiplicities are $\rho_0 = 1, \rho_1 = v_1 - 1, \rho_2 = (v_2 - 1)\, v_1 v_3, \rho_3 = (v_3 - 1)$ and $\rho_4 = (v_1 - 1)(v_3 - 1)$. The idempotent matrices can be easily obtained.

Right angular PBIB designs can be obtained from the generalized right angular designs by substituting $v_1 = 2$. The parameters of the derived designs are $2v_2v_3$, $b, r, k, \lambda_1', \lambda_2', \lambda_3', \lambda_4', n_1 = v_2 - 1, n_2 = v_2, n_3 = n_4 = v_2(v_3 - 1)$. The eigen values of $\mathbf{NN}'$ are

$$\theta_0 = rk,$$

$$\theta_1 = r - \lambda_1' + v_2[\lambda_1' - \lambda_2' + (v_3 - 1)(\lambda_3' - \lambda_4')],$$

$$\theta_2 = r - \lambda_1',$$

$$\theta_3 = r - \lambda_1' + v_2[\lambda_1' + \lambda_2' - \lambda_3' - \lambda_4']$$

and

$$\theta_4 = r - \lambda_1' + v_2[\lambda_1' - \lambda_2' - \lambda_3' + \lambda_4']$$

with multiplicities

$$\rho_0 = 1, \rho_1 = 1, \rho_2 = 2(v_2 - 1)\, v_3, \rho_3 = (v_3 - 1) \text{ and } \rho_4 = (v_3 - 1).$$

4.7. OTHER KRONECKER PRODUCT DESIGNS

We have seen in previous Sections that the Kronecker product of two or more incidence matrices of incomplete block designs leads to new designs. It was also shown that a linear combination of Kronecker products also leads to new designs. This was the case while obtaining family (A) BIB designs in Section 4.3.1. We now describe another linear combination of Kronecker products introduced by Nigam and Mathur (1974) and subsequently studied by Sinha, Mathur and Nigam (1979). They introduced an operation called Kronecker sum of two matrices, say, $\mathbf{A} = (a_{ij})$ and $\mathbf{B} = (b_{ij})$ of sizes $m \times n$ and $p \times q$ respectively as

$$\mathbf{A} \circledS \mathbf{B}$$

where $\circledS$ denotes Kronecker sum and is defined as

$$\mathbf{A} \circledS \mathbf{B} = \begin{bmatrix} a_{11}\mathbf{11}' + \mathbf{B} & \ldots & a_{1n}\mathbf{11}' + \mathbf{B} \\ a_{21}\mathbf{11}' + \mathbf{B} & \ldots & a_{2n}\mathbf{11}' + \mathbf{B} \\ & \ldots & \\ & \ldots & \\ a_{m1}\mathbf{11}' + \mathbf{B} & \ldots & a_{mn}\mathbf{11}' + \mathbf{B} \end{bmatrix}$$

Clearly $\mathbf{A} \circledS \mathbf{B}$ can be expressed as

$$\mathbf{A} \circledS \mathbf{B} = \mathbf{A} \otimes \mathbf{11}' + \mathbf{11}' \otimes \mathbf{B}$$

where $\mathbf{11}'$ is a matrix with all elements unity.

If **A** and **B** are (0, 1) matrices, then $\mathbf{A} \circledcirc \mathbf{B}$ is a matrix with entries 0, 1 and 2. The following transformations are true for such matrices.

(i) Transformation $2 \to 1, \quad 1 \to 0, \quad 0 \to 0$

$\mathbf{A} \circledcirc \mathbf{B} = \mathbf{A} \otimes \mathbf{B}$

(ii) Transformation $2 \to 0, \quad 1 \to 1, \quad 0 \to 0$

$\mathbf{A} \circledcirc \mathbf{B} = \bar{\mathbf{A}} \otimes \mathbf{B} + \mathbf{A} \otimes \bar{\mathbf{B}}$

where $\bar{\mathbf{A}}$ and $\bar{\mathbf{B}}$ are complementary matrices of **A** and **B** respectively.

(iii) Transformation $2 \to 1, \quad 1 \to 1, \quad 0 \to 0$

$\mathbf{A} \circledcirc \mathbf{B} = \bar{\mathbf{A}} \otimes \mathbf{B} + \mathbf{A} \otimes \mathbf{J}_{p \times q}$, where $\mathbf{J}_{p \times q}$ is a matrix with all elements unity.

(iv) Transformation $2 \to 1, \quad 1 \to 0, \quad 0 \to 1$

$\mathbf{A} \circledcirc \mathbf{B} = \bar{\mathbf{A}} \otimes \bar{\mathbf{B}} + \mathbf{A} \otimes \mathbf{B}$

It can be proved that if $\mathbf{N}_1$ is the incidence matrix of an m_1-associate PBIB design and $\mathbf{N}_2$ is the incidence matrix of an m_2-associate PBIB design then $\mathbf{N} = \mathbf{N}_1 \circledcirc \mathbf{N}_2$ is a partially balanced ternary design with $(m_1 + m_2 + m_1 m_2)$-associate classes. The ternary design reduces to a binary (hence PBIB) design under the different transformations (i) to (iv) listed above. Many useful PBIB designs can, therefore, be constructed through this procedure.

5. Block Designs with Factorial Structure

5.1 INTRODUCTION

In this Chapter the readers will be exposed to the interesting work on block designs with factorial structure. For such designs it is possible to develop a one-to-one correspondence between the treatments of the block designs and the treatment combinations of a factorial experiment. Such a correspondence helps in analysing block designs as confounded factorials and vice-versa. Some earlier attempts to identify such a correspondence can be traced back to the development of quasi-factorials or lattice designs. An account of these designs is available in text books (See e.g., Kempthorne, 1952).

We now describe an example to clarify the idea of correspondence between the block designs and factorial experiments. Consider a resolvable BIB design with blocks arranged as follows:

B_1	B_2	B_3	B_4	B_5	B_6
1	3	1	2	1	2
2	4	3	4	4	3

Let us associate with this design a 2^2-factorial experiment with two-factors as F_1 and F_2. Let the two levels of each of the two factors be denoted by 0 and 1. Assign the four treatment combinations for the factorial experiment to the four treatments of the BIB design through the following correspondence:

$$00 \to 1;\ 01 \to 2;\ 10 \to 3;\ 11 \to 4.$$

This gives the following factorial design:

		Blocks			
1	2	3	4	5	6
00	10	00	01	00	01
01	11	10	11	11	10

For the BIB design, the concurrence matrix is of the form

$$\mathbf{NN}' = \begin{bmatrix} 3 & 1 & 1 & 1 \\ 1 & 3 & 1 & 1 \\ 1 & 1 & 3 & 1 \\ 1 & 1 & 1 & 3 \end{bmatrix} \tag{5.1}$$

and $$\mathbf{M} = \mathbf{NN}'/rk = \mathbf{NN}'/6.$$

The main effects F_1 and F_2 and the interaction F_1F_2 can be estimated by using the following three mutually orthogonal observational contrasts:

$$\begin{array}{cccc} -(00) & -(01) & +(10) & +(11) \\ -(00) & +(01) & -(10) & +(11) \\ +(00) & -(01) & -(10) & +(11) \end{array}$$

where (00), (01), (10) and (11) denote, respectively, the total of the observations from the experimental units receiving the treatment combinations 00, 01, 10 and 11.

The vectors of co-efficients $\mathbf{s}_1$, $\mathbf{s}_2$ and $\mathbf{s}_3$, corresponding to these contrasts, can be written as

$$\begin{aligned} \mathbf{s}_1 &= (-1\ -1\ 1\ 1)' = (-1\ 1)' \otimes (1\ 1)' \\ \mathbf{s}_2 &= (-1\ 1\ -1\ 1)' = (1\ 1)' \otimes (-1\ 1)', \\ \mathbf{s}_3 &= (1 - 1 - 1\ 1)' = (-1\ 1)' \otimes (-1\ 1)'. \end{aligned} \tag{5.2}$$

It can be verified easily that these contrasts are eigen vectors of $\mathbf{NN}'$ and hence of $\mathbf{M}$. These contrasts are therefore basic contrasts. As the corresponding eigen values of $\mathbf{M}$ are each equal to 1/3, it follows that the main effects and the two-factor interaction are estimated with the same loss of information 1/3. We shall see later that every BIB design with $v = m_1 \times \ldots \times m_n$

treatments can always be used as an $m_1 \times \ldots \times m_n$ factorial experiment and for such a design every effect is estimated with same relative loss of information $(r-\lambda)/rk$. As a factorial experiment this class of designs has little utility because all the effects are confounded with the blocks to the same extent.

Consider now the following GD PBIB design with groups (00,01) and (10, 11) and the parameters $v=4$, $b=4$, $r=2$, $k=2$, $\lambda_1=0$, $\lambda_2=1$, $m=2$, $n=2$. The plan of the design can be obtained by deleting the first two blocks from the plan of the BIB design just considered.

Blocks

1	2	3	4
00	01	00	01
10	11	11	10

The concurrence matrix of this design is

$$\mathbf{NN}' = \begin{bmatrix} 2 & 0 & 1 & 1 \\ 0 & 2 & 1 & 1 \\ 1 & 1 & 2 & 0 \\ 1 & 1 & 0 & 2 \end{bmatrix} \tag{5.3}$$

and $$\mathbf{M} = \mathbf{NN}'/rk = \mathbf{NN}'/4.$$

It is easy to verify that $\mathbf{s}_1$ is an eigen vector of $\mathbf{NN}'$ corresponding to the eigen value 0 and $\mathbf{s}_2$ and $\mathbf{s}_3$ are the eigen vectors of $\mathbf{NN}'$ corresponding to the eigen value 2, where $\mathbf{s}_1$, $\mathbf{s}_2$ and $\mathbf{s}_3$ are as defined in (5.2). Thus the block design provides a factorial experiment in which the main-effect F_1 is unconfounded and is therefore estimated with full efficiency, whereas the main effect F_2 and the interaction F_1F_2 are each estimated with relative loss of information 1/2. As a design for a factorial experiment this is superior to BIB design as this design not only estimates F_1 free from block effects but also requires less number of experimental units.

Remark 5.1: Notice that the structures of $\mathbf{NN}'$ of (5.1) and (5.3) are cyclic in terms of partitioned matrices, i.e.

$$\mathbf{NN}' = \{\mathbf{N}_1 \ \ \mathbf{N}_2\},$$

where { } denotes the cyclic pattern. The $\mathbf{N}_1$ and $\mathbf{N}_2$ matrices in (5.1) and (5.3) are

$$\begin{bmatrix} 3 & 1 \\ 1 & 3 \end{bmatrix}, \begin{bmatrix} 1 & 1 \\ 1 & 1 \end{bmatrix} \text{ and } \begin{bmatrix} 2 & 0 \\ 0 & 2 \end{bmatrix}, \begin{bmatrix} 1 & 1 \\ 1 & 1 \end{bmatrix},$$

respectively. Also row sums and column sums of $\mathbf{N}_1$ and $\mathbf{N}_2$ matrices are constant. It will be seen later that such a structural form of $\mathbf{NN}'$ helps in simplifying the analysis.

Remark 5.2: The structure of the contrasts as defined in (5.2) reveals that the higher order contrasts can be expressed as the Kronecker product of lower dimensional contrasts or eigen vectors. This property, which is true in general, is extremely helpful in determining the pattern of analysis. It has already been seen in Chapter 4 that the treatment contrasts of Kronecker product designs have such patterns.

5.2. TREATMENT AND FACTORIAL EFFECTS

We now proceed to establish a relationship between the treatment effects of a block design and the factorial effects of the corresponding factorial experiment. The work is due to Kurkjian and Zelen (1962). Consider a factorial experiment with n factors $F_1, \ldots, F_n$, where the j-th factor has m_j levels, $j = 1, \ldots, n$. A particular selection of levels $i = (i_1, \ldots, i_n), i_j = 0, 1, \ldots, m_j - 1; j = 1, \ldots, n$, will be denoted as the i-th treatment combination, there being $v = \prod_{j=1}^{n} m_j$ such treatment combinations. Let $\boldsymbol{\theta}_j = (0, 1, \ldots, m_j - 1)'$ be a vector whose elements denote the levels at which the factor F_j appears. Then the Symbolic Direct Product (SDP) of any two $\boldsymbol{\theta}_j$'s is defined as

$$(\boldsymbol{\theta}_p \times \boldsymbol{\theta}_q) = (00, 01, \ldots, 0m_q-1, 10, \ldots, 1m_q-1, \ldots, \\ m_p-1, 0, \ldots, m_p-1m_q-1)'. \tag{5.4}$$

(It may be noted that the symbol '$\times$' represents the SDP throughout this Section although elsewhere this might represent multiplication.)

The SDP among all $\boldsymbol{\theta}_j$'s ($j = 1, \ldots, n$) can be represented as

$$\Phi_n = \boldsymbol{\theta}_1 \times \ldots \times \boldsymbol{\theta}_n = \prod_{j=1}^{n} \times \boldsymbol{\theta}_j. \tag{5.5}$$

The rows of Φ_n can be used to order lexicographically the v treatment combinations by designating the i-th treatment combination to the i-th row of Φ_n. The element of j-th column in the i-th row of Φ_n denotes the level of the factor F_j in the i-th treatment combination. Throughout, we shall assume the v treatments to be arranged lexicographically. We can now write any set of t ($\leqslant n$) columns of $\boldsymbol{\Phi}_n$ as

$$\boldsymbol{\theta}^X = \boldsymbol{\theta}_1^{X_1} * \ldots * \boldsymbol{\theta}_n^{X_n},$$

where $\boldsymbol{\theta}_j^{X_j} = \boldsymbol{\theta}_j$ if $X_j = 1$, i.e. if the j-th column is selected,

$= \mathbf{1}_{m_j}$ if $X_j = 0$, i.e., if the j-th column is not selected,

such that $\sum\limits_{j=1}^{n} X_j = t$ and the symbol $*$ is either the Kronecker Product or the SDP.

To illustrate this, consider a 3×2^2 factorial experiment. Here

$$\boldsymbol{\Phi}_3 = \boldsymbol{\theta}_1 \times \boldsymbol{\theta}_2 \times \boldsymbol{\theta}_3, \text{ i.e.}$$

$$\boldsymbol{\Phi}_3 = \begin{pmatrix} 0 \\ 1 \\ 2 \end{pmatrix} \times \begin{pmatrix} 0 \\ 1 \end{pmatrix} \times \begin{pmatrix} 0 \\ 1 \end{pmatrix}$$

$$= \begin{bmatrix} 0 & 0 & 0 & 0 & 1 & 1 & 1 & 1 & 2 & 2 & 2 & 2 \\ 0 & 0 & 1 & 1 & 0 & 0 & 1 & 1 & 0 & 0 & 1 & 1 \\ 0 & 1 & 0 & 1 & 0 & 1 & 0 & 1 & 0 & 1 & 0 & 1 \end{bmatrix}'. \tag{5.6}$$

The first column of $\boldsymbol{\Phi}_3$ is

$$\boldsymbol{\theta}_1 \otimes \mathbf{1}_2 \otimes \mathbf{1}_2 = \begin{pmatrix} 0 \\ 1 \\ 2 \end{pmatrix} \otimes \begin{pmatrix} 1 \\ 1 \end{pmatrix} \otimes \begin{pmatrix} 1 \\ 1 \end{pmatrix}$$

$$= (0 \quad 0 \quad 0 \quad 0 \quad 1 \quad 1 \quad 1 \quad 1 \quad 2 \quad 2 \quad 2 \quad 2)'.$$

Similarly, the second and third columns of $\boldsymbol{\Phi}_3$ are obtained as $\mathbf{1}_3 \otimes \boldsymbol{\theta}_2 \otimes \mathbf{1}_2$ and $\mathbf{1}_3 \otimes \mathbf{1}_2 \otimes \boldsymbol{\theta}_3$, respectively.

The first and third columns of $\boldsymbol{\Phi}_3$ taken together can be obtained as

$$\boldsymbol{\theta}_1 \times \mathbf{1}_2 \otimes \boldsymbol{\theta}_3 = \begin{bmatrix} 0 & 0 & 0 & 0 & 1 & 1 & 1 & 1 & 2 & 2 & 2 & 2 \\ 0 & 1 & 0 & 1 & 0 & 1 & 0 & 1 & 0 & 1 & 0 & 1 \end{bmatrix}'.$$

Consider now a factorial arrangement of v treatment combinations incorporated in a block design $D(v, b, r, k)$. We assume the usual intra block fixed effects model. The treatment effects in the block design can be written in terms of main-effect and interaction parameters. The i-th treatment effect can thus be represented as

$$\tau_i = \sum_{j=1}^{n} a_j(i_j) + \sum_{j<k}\sum a_{jk}(i_j, i_k) + \ldots + a_{1\cdots n}(i_1, \ldots, i_n), \quad (5.7)$$

where the i-th treatment effect corresponds to the i-th treatment combination, $(i_1,\ldots, i_n)$ say. Here $a_j(i_j)$ denotes the main-effect parameters of the j-th factor F_j at level i_j in the i-th treatment combination, $a_{jk}(i_j, i_k)$ denotes the effect of two factor interaction associated with factors F_j and F_k at levels i_j and i_k, respectively in the i-th treatment combination, and so on. The following linear constraints are satisfied:

$$\text{i)} \quad \sum_{i_j=0}^{m_j-1} a_j(i_j) = 0, \text{ for all } j = 1,\ldots, n,$$

$$\text{ii)} \quad \sum_{i_j=0}^{m_j-1} a_{jk}(i_j, i_k) = 0 = \sum_{i_k=0}^{m_k-1} a_{jk}(i_j, i_k), \quad (5.8)$$

and so on.

Consider once again the 3×2^2 factorial experiment. Write the treatment combinations as in (5.6). We then have

$$\tau_1 = a_1(0)+a_2(0)+a_3(0)+a_{12}(0, 0)+a_{13}(0, 0)+a_{23}(0, 0)+a_{123}(0, 0, 0)$$

$$\vdots$$

$$\tau_{12} = a_1(2)+a_2(1)+a_3(1)+a_{12}(2, 1)+a_{13}(2, 1)+a_{23}(1, 1)+a_{123}(2, 1, 1).$$

Remark 5.3: Notice that the first element of the above equations corresponds to the elements of $\mathbf{a}_1 \otimes \mathbf{1}_2 \otimes \mathbf{1}_2$, which is the same as the first column of Φ_3, obtained by replacing $\boldsymbol{\theta}_1$ by $\mathbf{a}_1$, where $\mathbf{a}_1 = (a_1(0), a_1(1), a_1(2))$. Similarly, all other elements of the above equations correspond to some column or columns of Φ_3.

We can now write $\boldsymbol{\tau}$ as

$$\boldsymbol{\tau} = \mathbf{a}_1 \otimes \mathbf{1}_2 \otimes \mathbf{1}_2 + \mathbf{1}_3 \otimes \mathbf{a}_2 \otimes \mathbf{1}_2 + \mathbf{1}_3 \otimes \mathbf{1}_2 \otimes \mathbf{a}_3 + \mathbf{a}_1 \times \mathbf{a}_2 \otimes \mathbf{1}_2$$
$$+ \mathbf{a}_1 \times \mathbf{1}_2 \otimes \mathbf{a}_3 + \mathbf{1}_3 \otimes \mathbf{a}_2 \times \mathbf{a}_3 + \mathbf{a}_1 \times \mathbf{a}_2 \times \mathbf{a}_3,$$

which, in a general form, can be expressed as

$$\boldsymbol{\tau} = \sum_{d=1}^{n} \{\sum_{X_1+\ldots+X_n=d} \mathbf{a}_1^{X_1} * \ldots * \mathbf{a}_n^{X_n}\}, \quad (5.9)$$
$$= \sum_{X(1,\, n)}' \mathbf{a}^{X},$$

where the abbreviated notation $\sum'_{X(s,\, n)}$ is used for the summation of n digit binary numbers $\mathbf{X} = (X_1,\ldots, X_n)$ for which sum of the elements of $\mathbf{X}$ is atleast equal to s. Here $X_j = 0$ or 1 depending upon whether a given factor is present or absent in the effect, and $\mathbf{a}_j^{X_j} = \mathbf{a}_j$ if $X_j = 1$ and $\mathbf{a}_j^{X_j} = \mathbf{1}_{m_j}$ if $X_j = 0$. A generalized interaction may therefore be represented as $\mathbf{a}^X$. The factors present in the interaction are those for which $X_j = 1$.

We now define $\mathbf{C}_j$ matrices of order m_j as

$$\mathbf{C}_j = \mathbf{I}_{m_j} - \mathbf{1}_{m_j}\mathbf{1}'_{m_j}/m_j. \tag{5.10}$$

These are contrast matrices as they satisfy $\mathbf{C}_j\,\mathbf{1}_{m_j} = \mathbf{0}$. Let us now define

$$\begin{aligned} \mathbf{C}_j^{X_j} &= \mathbf{C}_j \text{ if } X_j = 1, \\ &= \mathbf{1}'_{m_j}/m_j \text{ if } X_j = 0. \end{aligned} \tag{5.11}$$

We also have $\mathbf{1}'_{m_j}\mathbf{1}_{m_j} = m_j$, $\mathbf{C}_j\mathbf{a}_j = \mathbf{a}_j$ and $\mathbf{1}'_{m_j}\mathbf{a}_j = 0$.

Utilizing these relations we have

$$\begin{aligned} \mathbf{C}_j^{X_j}\mathbf{a}_j^{Y_j} &= 1 \text{ if } X_j = Y_j = 0, \\ &= \mathbf{a}_j \text{ if } X_j = Y_j = 1, \\ &= 0 \text{ otherwise.} \end{aligned} \tag{5.12}$$

Utilising (5.12), it is easy to show that for $\sum_{j=1}^{n} X_j = p$, we have

$$\begin{aligned} &= [\mathbf{C}_1^{X_1} \times \ldots \times \mathbf{C}_n^{X_n}]\,\boldsymbol{\tau} \\ &= \sum_{d=1}^{n} \{\sum_{X_1+\ldots+X_n=d} \mathbf{C}_1^{X_1}\,\mathbf{a}_1^{Y_1} * \ldots * \mathbf{C}_n^{X_n}\mathbf{a}_n^{Y_n}\} \\ &= \mathbf{a}_{j_1} \times \ldots \times \mathbf{a}_{j_p}, \end{aligned}$$

because for a given vector $\mathbf{X}$ there exists only one vector $\mathbf{Y}$ satisfying $\mathbf{Y}=\mathbf{X}$ and $\sum_{j=1}^{n} X_j = \sum_{j=1}^{n} Y_j = p$. Hence the p-factor interaction $F_{j_1} \ldots F_{j_p}$ will be

$$\mathbf{a}_{j_1} \times \ldots \times \mathbf{a}_{j_p} = [\mathbf{C}_1^{X_1} \otimes \ldots \otimes \mathbf{C}_n^{X_n}]\boldsymbol{\tau}. \tag{5.13}$$

If we write

$$\mathbf{C}^X = \prod_{j=1}^{n} \otimes\, \mathbf{C}_j^{X_j}, \tag{5.14}$$

then (5.13) can be written as

$$\mathbf{a}^X = \mathbf{C}^X\boldsymbol{\tau}. \tag{5.15}$$

The expression (5.15) is important as it connects the parameters of the block design with the parameters of the corresponding factorial experiment.

For the 3×2^2 factorial experiment the main-effects and interactions can be derived from the following:

$$\begin{aligned} \mathbf{a}_1 &= (\mathbf{C}_1 \otimes \mathbf{1}_2' \otimes \mathbf{1}_2')\,\boldsymbol{\tau} \\ \mathbf{a}_2 &= (\mathbf{1}_3' \otimes \mathbf{C}_2 \otimes \mathbf{1}_2')\,\boldsymbol{\tau} \\ \mathbf{a}_3 &= (\mathbf{1}_3' \otimes \mathbf{1}_2' \otimes \mathbf{C}_3)\,\boldsymbol{\tau} \end{aligned}$$

$$\mathbf{a}_1 \times \mathbf{a}_2 = (\mathbf{C}_1 \otimes \mathbf{C}_2 \otimes \mathbf{1}'_2)\, \boldsymbol{\tau}$$

$$\mathbf{a}_1 \times \mathbf{a}_3 = (\mathbf{C}_1 \otimes \mathbf{1}'_2 \otimes \mathbf{C}_3)\, \boldsymbol{\tau}$$

$$\mathbf{a}_2 \times \mathbf{a}_3 = (\mathbf{1}'_3 \otimes \mathbf{C}_2 \otimes \mathbf{C}_3)\, \boldsymbol{\tau}$$

$$\mathbf{a}_1 \times \mathbf{a}_2 \times \mathbf{a}_3 = (\mathbf{C}_1 \otimes \mathbf{C}_2 \otimes \mathbf{C}_3)\, \boldsymbol{\tau}.$$

5.3 CYCLIC STRUCTURE OF Ω-MATRIX AND FACTORIAL EXPERIMENTS

It has been shown in Section 5.2 that the factorial effects can be expressed in terms of treatment effects. We now proceed to obtain conditions under which block designs have factorial structure.

Definition 5.1: A block design is said to have an orthogonal factorial structure if the adjusted sum of squares due to treatments in the block design can be partitioned orthogonally into sum of squares corresponding to main-effects and interactions in a factorial experiment.

John and Smith (1972) obtained a sufficient condition on $\boldsymbol{\Omega}$ matrix of a block design to have a factorial structure. They considered the case of two factor experiment. Their results were generalized by Cotter, John and Smith (1973) to the case of n factors.

Recalling the connection between the factorial effects and block design treatment parameters, as given in (5.15), we have the estimates of the factorial effects as

$$\hat{\mathbf{a}}^X = \mathbf{C}^X \hat{\boldsymbol{\tau}} \tag{5.16}$$

where $\hat{\boldsymbol{\tau}} = \boldsymbol{\Omega}\mathbf{Q}$. The dispersion matrix of the estimates $\hat{\mathbf{a}}^X$ is given by

$$D\,(\hat{\mathbf{a}}^X) = \sigma^2\, \mathbf{C}^X \boldsymbol{\Omega} \mathbf{C}^{X'}, \tag{5.17}$$

and

$$Cov\,(\hat{\mathbf{a}}^X, \hat{\mathbf{a}}^Y) = \sigma^2\, \mathbf{C}^X \boldsymbol{\Omega} \mathbf{C}^{Y'}. \tag{5.18}$$

The adjusted sum of squares for testing the significance of any generalized interaction $\mathbf{a}^X$ is given by

$$(\mathbf{C}^X \hat{\boldsymbol{\tau}})'\, (\mathbf{C}^X \boldsymbol{\Omega} \mathbf{C}^{X'})^-\, \mathbf{C}^X \hat{\boldsymbol{\tau}}. \tag{5.19}$$

In view of Definition 5.1 a design will have factorial structure if and only if $\mathbf{C}^X \boldsymbol{\Omega} \mathbf{C}^{Y'} = \mathbf{Q}$, for all $X \neq Y$.

John and Smith (1972) considered the case of two factors F_1 and F_2 at m_1 and m_2 levels, respectively. They assumed a cyclic pattern of $\boldsymbol{\Omega}$ matrix in respect of partitioned matrices $\boldsymbol{\Omega}_i$ for the resulting design to have an orthogonal factorial structure, i.e.

$$\boldsymbol{\Omega} = \begin{bmatrix} \boldsymbol{\Omega}_1 & \boldsymbol{\Omega}_2 & \ldots\boldsymbol{\Omega}_{m_1} \\ \boldsymbol{\Omega}_{m_1} & \boldsymbol{\Omega}_1 & \ldots\boldsymbol{\Omega}_{m_1-1} \\ \vdots & & \\ \boldsymbol{\Omega}_2 & \boldsymbol{\Omega}_3 & \ldots\boldsymbol{\Omega}_1 \end{bmatrix} \tag{5.20}$$

or

$$\boldsymbol{\Omega} = \{\boldsymbol{\Omega}_1, \ \ldots, \ \boldsymbol{\Omega}_{m_1}\},$$

where the partitioned matrices $\boldsymbol{\Omega}_i$ $(i = 1, \ldots, m_1)$ are of order $m_2 \times m_2$.

Let $\mathbf{R}^i_{m_1}$ be a circulant matrix of order $m_1 \times m_1$. The first row of $\mathbf{R}^i_{m_1}$ has a unity in its i-th column and zero elsewhere, $i = 1, \ \ldots, m_1$. Then $\boldsymbol{\Omega}$ in (5.20) can alternatively be written as

$$\boldsymbol{\Omega} = \sum_i \mathbf{R}^i_{m_1} \otimes \boldsymbol{\Omega}_i. \tag{5.21}$$

Assume further that the row and column sums of the sub-matrices $\boldsymbol{\Omega}_i$ are constant, i.e. $\boldsymbol{\Omega}_i \mathbf{1}_{m_2} = w_i \mathbf{1}$. The contrast matrices in case of two factor factorial experiment can be written as

$$\mathbf{C}_{10} = \mathbf{C}_1 \otimes \mathbf{1}'_{m_2}/m_2,$$
$$\mathbf{C}_{01} = \mathbf{1}'_{m_1}/m_1 \otimes \mathbf{C}_2,$$
$$\mathbf{C}_{11} = \mathbf{C}_1 \otimes \mathbf{C}_2.$$

These matrices can be expressed as

$$\begin{aligned} m_1m_2\mathbf{C}_{10} &= m_1\mathbf{I}_{m_1} \otimes \mathbf{1}'_{m_2} - (\mathbf{1}_{m_1}\mathbf{1}'_{m_1}) \otimes \mathbf{1}'_{m_2} \\ &= m_1\mathbf{P}_1 - \mathbf{1}_{m_1}\mathbf{1}_v', \end{aligned}$$

where

$$\mathbf{P}_1 = \{\mathbf{1}'_{m_2}, \mathbf{0}, \ldots, \mathbf{0}\}. \tag{5.22}$$

Similarly,

$$m_1m_2\mathbf{C}_{01} = m_2\mathbf{P}_2 - \mathbf{1}_{m_2}\mathbf{1}_v'$$

and

$$\begin{aligned} m_1m_2\mathbf{C}_{11} = {} & m_1m_2 \{\mathbf{I}_{m_2}, \mathbf{0}, \ldots, \mathbf{0}\} - m_1 \{\mathbf{1}_{m_2}\mathbf{1}'_{m_2}, \mathbf{0}, \ldots, \mathbf{0}\} \\ & - m_2 \{\mathbf{I}_{m_2}, \ldots, \mathbf{I}_{m_2}\} + \{\mathbf{1}_{m_2}\mathbf{1}'_{m_2}, \ldots, \mathbf{1}_{m_2}\mathbf{1}'_{m_2}\}, \end{aligned} \tag{5.23}$$

where

$$\mathbf{P}_2 = \{\mathbf{I}_{m_2}, \ldots, \mathbf{I}_{m_2}\}.$$

Also, using (5.20) notice that

$$\begin{aligned} m_1m_2\mathbf{C}_{10}\boldsymbol{\Omega} &= m_1\mathbf{P}_1\boldsymbol{\Omega} - \mathbf{1}_{m_1}\mathbf{1}_v'\boldsymbol{\Omega} \\ &= m_1 \{\mathbf{1}'_{m_2}\boldsymbol{\Omega}_1, \ldots, \mathbf{1}'_{m_2}\boldsymbol{\Omega}_{m_1}\} - W\mathbf{1}_{m_1}\mathbf{1}_v' \\ &= m_1 \{W_1\mathbf{1}'_{m_2}, \ldots, W_{m_1}\mathbf{1}'_{m_2}\} - W\mathbf{1}_{m_1}\mathbf{1}_v' \\ &= m_1 \{\boldsymbol{\alpha}'_1, \ldots, \boldsymbol{\alpha}'_{m_1}\} - W\mathbf{1}_{m_1}\mathbf{1}_v', \end{aligned}$$

where $\quad \boldsymbol{\alpha}'_i = W_i\mathbf{1}'_{m_2}$ and $W = \sum_{i=1}^{m_1} W_i$.

As $\mathbf{C}_j\mathbf{1}_{m_j} = \mathbf{0}$ ($j = 1, 2, 3$), we have

$$\begin{aligned}(m_1m_2)^2\mathbf{C}_{10}\boldsymbol{\Omega}\mathbf{C}_{01}' &= m_1\{\boldsymbol{\alpha}_1',\ldots,\boldsymbol{\alpha}_{m_1}'\}(m_2\mathbf{P}_2 - \mathbf{1}_{m_2}\mathbf{1}_v')' \\ &= m_1m_2\{\boldsymbol{\alpha}_1',\ldots,\boldsymbol{\alpha}_{m_1}'\}(\mathbf{I}_{m_2},\ldots,\mathbf{I}_{m_2})' \\ &\quad - m_1\{\boldsymbol{\alpha}_1',\ldots,\boldsymbol{\alpha}_{m_1}'\}\mathbf{1}_v\mathbf{1}_{m_2}', \\ &= m_1m_2W\mathbf{1}_{m_1}\mathbf{1}_{m_2}' - m_1m_2W\mathbf{1}_{m_1}\mathbf{1}_{m_2}' \\ &= \mathbf{0}.\end{aligned}$$

Similarly, it can be shown that

$$\begin{aligned}(m_1m_2)^2\,\mathbf{C}_{10}\boldsymbol{\Omega}\mathbf{C}_{11}' &= m_1^2m_2\{\boldsymbol{\alpha}_1',\ldots,\boldsymbol{\alpha}_{m_1}'\} - m_1^2m_2\{\boldsymbol{\alpha}_1',\ldots,\boldsymbol{\alpha}_{m_1}'\} \\ &\quad + m_1m_2W\mathbf{1}_v\mathbf{1}_v' - m_1m_2W\mathbf{1}_v\mathbf{1}_v' \\ &= \mathbf{0}.\end{aligned}$$

Finally, it can be shown that

$$m_1m_2\,\mathbf{C}_{01}\boldsymbol{\Omega} = m_2\,\Big(\sum_{i=1}^{m_1}\boldsymbol{\Omega}_i,\ldots,\sum_{i=1}^{m_1}\boldsymbol{\Omega}_i\Big) - W\mathbf{1}_{m_2}\mathbf{1}_v',$$

and post multiplying by $\mathbf{C}_{11}'$ it follows that

$$\mathbf{C}_{01}\boldsymbol{\Omega}\mathbf{C}_{11}' = \mathbf{0}.$$

We can summarize these results in the form of the following theorem:

Theorem 5.1: For a two factor $m_1 \times m_2$ factorial experiment the design has factorial structure if
(i) its $\boldsymbol{\Omega}$ matrix can be partitioned as

$$\boldsymbol{\Omega} = \{\boldsymbol{\Omega}_1,\ldots,\boldsymbol{\Omega}_{m_1}\} = \{(\boldsymbol{\Omega}_i)\},$$

or equivalently, its $\boldsymbol{\Omega}$ matrix can be expressed as

$$\boldsymbol{\Omega} = \sum_{i=1}^{m_1}\mathbf{R}_{m_1}^i \otimes \boldsymbol{\Omega}_i,$$

and (ii) the row sums and column sums of $\boldsymbol{\Omega}_i$ matrices are constant, where $\boldsymbol{\Omega}_i$ is an $m_2 \times m_2$ matrix, $i = 1,\ldots, m_1$.

Cotter, John and Smith (1973) generalized the result of John and Smith (1972) to several factors. Their sufficient conditions are that the sub-matrices $\boldsymbol{\Omega}_i$ must have cyclic pattern and their row and column sums must be same. For an $m_1\times\ldots\times m_n$ factorial experiment the design will have factorial structure if $\boldsymbol{\Omega}$ matrix can be expressed as

$$\boldsymbol{\Omega} = \sum_{i_1=1}^{m_1} \sum_{i_2=1}^{m_2} \ldots \sum_{i_p=1}^{m_p} (\prod_{j=1}^{p} \otimes \mathbf{R}_{mj}^{ij}) \otimes \boldsymbol{\Omega}_{i_1,\ldots,i_p}, \tag{5.23a}$$

where $p = n - 1$ and $\boldsymbol{\Omega}_{i_1,\ldots,i_p}$ is an $m_n \times m_n$ matrix with all row sums and column sums constant.

For instance, in a three factor experiment with levels m_1, m_2 and m_3, the design will have a factorial structure if $\boldsymbol{\Omega}$ can be partitioned as

$$\boldsymbol{\Omega} = \{\boldsymbol{\Omega}_1, \ldots, \boldsymbol{\Omega}_{m_1}\}$$
$$= \{(\boldsymbol{\Omega}_{i_1})\},\ i_1 = 1, \ldots, m_1$$

and

$$\boldsymbol{\Omega}_{i_1} = \{\boldsymbol{\Omega}_{i_1,1}, \ldots, \boldsymbol{\Omega}_{i_1,m_2}\}$$
$$= \{(\boldsymbol{\Omega}_{i_1,i_2})\},$$

i.e. the $\boldsymbol{\Omega}$ matrix can be written as

$$\boldsymbol{\Omega} = \sum_{i_1=1}^{m_1} \sum_{i_2=1}^{m_2} \{\prod_{j=1}^{2} \otimes \mathbf{R}_{mj}^{ij}\} \otimes \boldsymbol{\Omega}_{i_1,i_2},$$

where $\boldsymbol{\Omega}_{i_1,i_2}$ are square circulant matrices of order $m_3 \times m_3$ such that the row sums and column sums of $\boldsymbol{\Omega}_{i_1,i_2}$ are constant.

Cotter, John and Smith have further shown that the adjusted sum of squares due to generalized interaction $\mathbf{a}^X$ given by (5.19) can be written as

$$\hat{\mathbf{a}}^X \mathbf{P}^X \mathbf{A} \mathbf{P}^{X'} \hat{\mathbf{a}}^{X'}, \tag{5.24}$$

where

$$\mathbf{P}^X = \prod_{j=1}^{n} \otimes \mathbf{P}_j$$

with

$$\mathbf{P}_j^{Xj} = \mathbf{I}_{mj} \text{ if } X_j = 1,$$
$$= \mathbf{1}'_{mj} \text{ if } X_j = 0,$$

and

$$\mathbf{A} = r\mathbf{I} - \mathbf{NN}'/k.$$

John (1981) has further simplified the expression (5.24) by noting that the treatment structure of a factorial experiment can be expressed by the identity

$$\boldsymbol{\tau} = \sum_X \mathbf{L}^X \boldsymbol{\tau} \tag{5.25}$$

where the summation is over all binary numbers $X = (X_1, \ldots, X_n)$, with $X_j = 0$ or 1 and

$$\mathbf{L}^X = \prod_{j=1}^{n} \otimes \mathbf{L}_j^{X_j} \tag{5.26}$$

with

$$\mathbf{L}_j^{X_j} = \mathbf{I}_j - \mathbf{1}_{m_j}\mathbf{1}'_{m_j}/m_j \text{ if } X_j = 1,$$

$$= \mathbf{1}_{m_j}\mathbf{1}'_{m_j}/m_j \text{ if } X_j = 0.$$

$\mathbf{L}^X$ matrices are mutually orthogonal idempotent matrices. We have already dealt with such matrices in previous Chapters.

Any generalized interaction $\mathbf{L}^X\boldsymbol{\tau}$, assumed to be estimable, is estimated by $\mathbf{L}^X\hat{\boldsymbol{\tau}}$ with the dispersion matrix

$$D(\mathbf{L}^X \hat{\boldsymbol{\tau}}) = \mathbf{L}^X\boldsymbol{\Omega}\mathbf{L}^X\sigma^2. \tag{5.27}$$

The readers can easily prove the following results:

$$\begin{aligned} \mathbf{L}^X\boldsymbol{\Omega}\mathbf{L}^Y &= \mathbf{L}^X\boldsymbol{\Omega} && \text{if } X = Y \\ &= \mathbf{0} && \text{if } X \neq Y \end{aligned} \tag{5.28}$$

and

$$\mathbf{L}^X\boldsymbol{\Omega} = \boldsymbol{\Omega}\mathbf{L}^X. \tag{5.29}$$

Again, as $\mathbf{L}^X\boldsymbol{\tau}$ is assumed to be estimable, we have from the estimability condition

$$\mathbf{L}^X = \mathbf{L}^X\boldsymbol{\Omega}\mathbf{A}. \tag{5.30}$$

The sum of squares due to the generalized interaction $\mathbf{L}^X\boldsymbol{\tau}$ (for testing H_0: $\mathbf{L}^X\boldsymbol{\tau} = 0$) can now be written as

$$(\mathbf{L}^X\hat{\boldsymbol{\tau}})' (\mathbf{L}^X\boldsymbol{\Omega}\mathbf{L}^X)^- \mathbf{L}^X\hat{\boldsymbol{\tau}}. \tag{5.31}$$

The expression (5.31) can be written as

$$\hat{\boldsymbol{\tau}}'\mathbf{L}^X\mathbf{A}\mathbf{L}^X\hat{\boldsymbol{\tau}},$$

which can further be simplified as

$$(\mathbf{L}^X\mathbf{Q})' (\mathbf{L}^X\hat{\boldsymbol{\tau}}). \tag{5.32}$$

The expression (5.32) is a considerable simplification on the expression (5.24).

Consider now a contrast in the treatment effects, $\mathbf{s}'\boldsymbol{\tau}$. Clearly it is a basic contrast if $\mathbf{s}$ is an eigen vector of $\boldsymbol{\Omega}$. A basic contrast in a generalized interaction can be represented as

$$\mathbf{s}^X = \prod_{j=1}^{n} \otimes \mathbf{s}^{X_j}, \tag{5.33}$$

where $m_j \times 1$ vector $\mathbf{s}_j^{X_j}$ is a contrast vector if $X_j = 1$ and a vector of all elements unity if $X_j = 0$. The contrasts at (5.2) are essentially of the same form as in (5.33). Since all the factorial effects are assumed to be estimable the eigen value h corresponding to the eigen vector $\mathbf{s}$ in (5.33) will be non-zero. Noting that $\mathbf{\Omega}\ \mathbf{s} = h\mathbf{s}$ and the dispersion matrix $D\,(\mathbf{s}'\hat{\boldsymbol{\tau}}) = \mathbf{s}'\mathbf{\Omega}\mathbf{s}\sigma^2 = h\mathbf{s}'\mathbf{s}\sigma^2$, the sum of squares due to $\mathbf{s}'\boldsymbol{\tau}$ is given by

$$(\mathbf{s}'\hat{\boldsymbol{\tau}})^2/h\,(\mathbf{s}'\mathbf{s}),$$

which can be rewritten as

$$h\,(\mathbf{s}'\mathbf{Q})^2/(\mathbf{s}'\mathbf{s}) \tag{5.34}$$

because $\mathbf{s}'\hat{\boldsymbol{\tau}} = \mathbf{s}'\mathbf{\Omega}\mathbf{Q} = h\mathbf{s}'\mathbf{Q}$. The expression (5.34) involves only the vector of adjusted treatment totals and the eigen value h. This expression is fairly easy to compute as h^{-1} is the eigen value of $\mathbf{A}$. Since the eigen values of $\mathbf{A}$ and $\mathbf{NN}'$ are also related in case of equi-replicate and proper designs, it is extremely simple to compute the sum of squares due to a generalized interaction.

From (5.26) and (5.33) it follows that $\mathbf{s}'\mathbf{L}^X = \mathbf{s}'$, so that

$$\mathbf{s}'\hat{\boldsymbol{\tau}} = h\mathbf{s}'\mathbf{Q} = h\mathbf{s}'\mathbf{L}^X\mathbf{Q}.$$

We therefore obtain an alternative expression for the sum of squares due to $\mathbf{s}'\boldsymbol{\tau}$ as

$$(\mathbf{s}'\mathbf{L}^X\mathbf{Q})'\,(\mathbf{s}'\mathbf{L}^X\hat{\boldsymbol{\tau}})/(\mathbf{s}'\mathbf{s}). \tag{5.35}$$

Since the eigen value corresponding to any eigen vector $\mathbf{s}$ of the form (5.33) is non-zero, it is possible to find an orthogonal decomposition of the factorial sum of squares in (5.32) into single degree of freedom components. The decomposition will be unique if for a given effect the eigen values corresponding to the set of eigen vectors spanning the contrast space are all distinct. If for a given interaction all eigen values are equal, then any orthogonal decomposition of (5.32) is possible. The design is then balanced with respect to this interaction. If a design is balanced with respect to all main-effects and interactions it will be said to have factorial balance.

The above results reveal that block designs with factorial structure admit a simple analysis and the covariance matrix $\mathbf{\Omega}$ is completely determined in terms of lower dimensional matrices $\mathbf{\Omega}_{i_1,i_2,\ldots,i_p}$ which are easier to compute. The pattern of analysis of these designs is also easily identified because of the structural form of $\mathbf{\Omega}$. It has already been discussed in Chapter 2 that in case of equireplicate and proper designs, $\mathbf{\Omega}$ and $\mathbf{NN}'$ have the same structural forms. It therefore follows that the pattern and analysis of block designs with factorial structure is completely determined by the concurrence matrix $\mathbf{NN}'$.

We have noticed that the block designs with factorial structure provide orthogonal estimates of various factorial effects. These designs can be factorially balanced or otherwise. We discuss the class of factorially balanced designs separately in Section 5.4. BIB designs and many of the known PBIB designs have factorial structure and therefore form a class of block designs useful for factorial experimentation. Cornish (1938) considered BIB designs as factorials whereas Harshbarger (1954), Brenna (1958) and Brenna and Kramer (1961) used factorials in Lattice designs. Kramer and Bradley (1957a, b) and Zelen (1958) extended these ideas further to use GD PBIB as factorials and Bradley, Walpole and Kramer (1960) examined two associate PBIB designs and showed that besides GD, PBIB designs based upon latin squares association scheme are also useful as factorial designs. Besides the BIB and PBIB designs many cyclic and generalized cyclic designs also form an important sub-class of designs with factorial structure. In view of their importance these are discussed separately in Chapter 6.

For PBIB designs, $\mathbf{NN}'$ and $\mathbf{\Omega}$ matrices have the same structure and therefore one can examine their factorial structure without inverting $\mathbf{\Omega}^{-1}$ matrix. We now consider some examples to clarify the ideas explained in this Section.

Consider the PBIB design listed as LS 18 in Clatworthy (1973). The design has the parameters $v = 16, r = 3, k = 3, b = 16, n_1 = 9, n_2 = 6, \lambda_1 = 0, \lambda_2 = 1$, with block contents as

10	7	16	7	9	4
11	8	13	8	10	1
12	5	14	5	11	2
9	6	15	6	12	3
13	4	6	2	16	9
14	1	7	3	13	10
15	2	8	4	14	11
16	3	5	1	15	12

The concurrence matrix **NN′** is given by

$$
\mathbf{NN'} = \left[\begin{array}{cccc|cccc|cccc|cccc}
3&0&0&0&0&0&1&1&0&1&0&1&0&1&1&0\\
0&3&0&0&1&0&0&1&1&0&1&0&0&0&1&1\\
0&0&3&0&1&1&0&0&0&1&0&1&1&0&0&1\\
0&0&0&3&0&1&1&0&1&0&1&0&1&1&0&0\\
0&1&1&0&3&0&0&0&0&0&1&1&0&1&0&1\\
0&0&1&1&0&3&0&0&1&0&0&1&1&0&1&0\\
1&0&0&1&0&0&3&0&1&1&0&0&0&1&0&1\\
1&1&0&0&0&0&0&3&0&1&1&0&1&0&1&0\\
0&1&0&1&0&1&1&0&3&0&0&0&0&0&1&1\\
1&0&1&0&0&0&1&1&0&3&0&0&1&0&0&1\\
0&1&0&1&1&0&0&1&0&0&3&0&1&1&0&0\\
1&0&1&0&1&1&0&0&0&0&0&3&0&1&1&0\\
0&0&1&1&0&1&0&1&0&1&1&0&3&0&0&0\\
1&0&0&1&1&0&1&0&0&0&1&1&0&3&0&0\\
1&1&0&0&0&1&0&1&1&0&0&1&0&0&3&0\\
0&1&1&0&1&0&1&0&1&1&0&0&0&0&0&3
\end{array}\right]
$$

The matrix **NN′** can be re-written in the partitioned form as

$$
\mathbf{NN'} = \{\mathbf{N_1}, \mathbf{N_2}, \mathbf{N_3}, \mathbf{N_4}\},
$$

where

$$
\mathbf{N_1} = \begin{bmatrix} 3&0&0&0\\ 0&3&0&0\\ 0&0&3&0\\ 0&0&0&3 \end{bmatrix} \qquad
\mathbf{N_2} = \begin{bmatrix} 0&0&1&1\\ 1&0&0&1\\ 1&1&0&0\\ 0&1&1&0 \end{bmatrix}
$$

$$\mathbf{N}_3 = \begin{bmatrix} 0 & 1 & 0 & 1 \\ 1 & 0 & 1 & 0 \\ 0 & 1 & 0 & 1 \\ 1 & 0 & 1 & 0 \end{bmatrix} \qquad \mathbf{N}_4 = \begin{bmatrix} 0 & 1 & 1 & 0 \\ 0 & 0 & 1 & 1 \\ 1 & 0 & 0 & 1 \\ 1 & 1 & 0 & 0 \end{bmatrix}$$

The sub matrices $\mathbf{N}_1$, $\mathbf{N}_2$, $\mathbf{N}_3$ and $\mathbf{N}_4$ are cyclic with row and column totals equal. Their pattern clearly satisfies the John and Smith (1972) criterion and therefore this block design is suitable for 4×4 two-factor factorial experiment.

The $\mathbf{NN}'$ matrix can also be written as

$$\mathbf{NN}' = \{\mathbf{N}_1, \mathbf{N}_2\}$$

where $\mathbf{N}_1$ and $\mathbf{N}_2$ are 8×8 matrices with row (and column) totals 5 (and 4), respectively. Thus the design is also suitable for 2×8 factorial experiment. Again, the 8×8 sub-matrices $\mathbf{N}_1$ and $\mathbf{N}_2$ can be further re-written in the cyclic partitioned form as $\mathbf{N}_1 = \{\mathbf{N}_{11}, \mathbf{N}_{12}\}$, $\mathbf{N}_2 = \{\mathbf{N}_{21}, \mathbf{N}_{22}\}$, where $\mathbf{N}_{11}$, $\mathbf{N}_{12}$ and $\mathbf{N}_{21}$, $\mathbf{N}_{22}$ are 4×4 matrices with constant row and column sums. The block design, therefore, is also suitable for $2^2 \times 4$ factorial experiment. No further cyclic partitioning with row and column totals equal is possible. Therefore, the block design when taken as, say, 2^4 factorial experiment does not have a factorial structure.

Cotter, John and Smith (1973) have shown that LS 67 of Clatworthy (1973) with parameters $v = 36$, $r = 10$, $k = 5$, $b = 72$, $n_1 = 10$, $n_2 = 5$, $\lambda_1 = 4$, $\lambda_2 = 0$, has factorial structures for 2×18, 3×12, 6×6, $2 \times 3 \times 6$ experiments, but the 4×9 and $2 \times 3 \times 3 \times 2$ experiments obtained through the same block design do not have a factorial structure.

Das (1960a) has also reported designs having factorial structure. For instance, for a $3^2 \times 2$ experiment in 6 plot blocks, Das has reported the following design in 4 replicates:

Replication I			*Replication II*		
B_1	B_2	B_3	B_4	B_5	B_6
100	200	000	200	000	100
010	110	210	110	210	010
220	020	120	020	120	220
201	001	101	101	201	001
111	211	011	011	111	211
021	121	221	221	021	121

Replication III			*Replication IV*		
B_7	B_8	B_9	B_{10}	B_{11}	B_{12}
100	200	000	200	000	100
210	010	110	010	110	210
020	120	220	120	220	020
201	001	101	101	201	001
011	111	211	211	011	111
121	221	021	021	121	221

If we consider the design with first or last two replicates, then it is easy to verify that these have factorial structures. The concurrence matrices of these designs can be written as $\{\mathbf{N}_1, \mathbf{N}_2, \mathbf{N}_3\}$ and $\{\mathbf{N}_1, \mathbf{N}_3, \mathbf{N}_2\}$ respectively, with partitioned matrices of order 6×6, where $\mathbf{N}_1 = \{\mathbf{N}_{11}, \mathbf{N}_{12}, \mathbf{N}_{12}\}$, $\mathbf{N}_2 = \{\mathbf{N}_{12}, \mathbf{N}_{12}, \mathbf{N}_{11}\}$ and $\mathbf{N}_3 = \{\mathbf{N}_{12}, \mathbf{N}_{11}, \mathbf{N}_{12}\}$, with sub-matrices of order 2×2 as $\mathbf{N}_{11} = 2\mathbf{I}$ and $\mathbf{N}_{12} = \mathbf{1}_2\mathbf{1}_2' - \mathbf{I}$.

The concurrence matrices clearly show that both the designs have factorial structure for $3^2\times2$ factorial experiment. Clearly, two more designs for factorial experiments viz. 2×9 and 3×6 with factorial structures can be obtained from the above designs.

5.4 FACTORIALLYB ALANCED DESIGN

Balancing is a desirable property of any experimental design. When the factorial structure is superimposed on the block designs then the balancing must be defined with respect to main effects and interactions of the factors. Bose (1947*b*) considered this for symmetrical factorial experiments in which each factor has the same number of levels. Shah (1958) and Puri and Nigam (1976) extended these considerations to asymmetrical factorial experiments. We now define the complete balance due to Shah (1958, 1960).

Definition 5.2: Complete balance is achieved over an interaction if and only if all the normalized contrasts belonging to the same interaction are estimated with the same variance.

Puri and Nigam (1976) gave an alternative definition of complete balance. According to their definition, complete balance is achieved over an interaction if the relative loss of information for every contrast belonging to that interaction is same, i.e. all the $\prod_{j=1}^{n} (m_j - 1)^{x_j}$ contrasts belonging to that interaction are estimated with same relative loss of information, say $\mu_{x_1 \ldots x_n}$. Their definition, however, coincides with that of Shah (1958) for equireplicate designs.

Shah (1958, 1960) defined a balanced factorial experiment (BFE) as follows:

Definition 5.3: A factorial experiment will be called a balanced factorial experiment if the following conditions are satisfied:

i) Each treatment is replicated the same number of times.
ii) Each of the block has the same number of plots.
iii) Estimates of the contrasts belonging to different interactions are uncorrelated with each other.
iv) Complete balance is achieved over each of the interactions.

BFE are also discussed in the literature under the nomenclature of factorial balance. We shall, however, refer to these as factorially balanced designs. Shah (1958) and Kshirsagar (1966) proved that BFE are PBIB designs with EGD association scheme with at most $2^n - 1$ associate classes. Kurkjian and Zelen (1963) introduced a structural property of **NN**′, called property (A). They showed that designs with property (A) satisfy the above requirements and hence are balanced factorials. They proved the following theorem:

Theorem 5.2: Let $m_j \times m_j$ matrices $\mathbf{L}_j^{X_j}$ be as defined in (5.26) and let the constants $h_{X_1 \ldots X_n}$ be defined by a function of the X's. Then if $D(\hat{\boldsymbol{\tau}})$ can be written in the form

$$D(\hat{\boldsymbol{\tau}}) = \sigma^2 \sum_{d=1}^{n} \{\sum_{X_1+\ldots+X_n=d} h_{X_1\ldots X_n} \prod_{j=1}^{n} \otimes \mathbf{L}_j^{X_j}\}$$

$$= \sigma^2 \sum_{X(1,n)}{}' \; h_X \mathbf{L}^X \tag{5.36}$$

then we shall have

i) $$V(\hat{\mathbf{a}}_{j_1} \times \ldots \times \hat{\mathbf{a}}_{j_p}) = \sigma^2 \prod_{j=1}^{n} m_j^{X_j} (\mathbf{C}_{j_1} \otimes \ldots \otimes \mathbf{C}_{j_p})/rvE_X, \tag{5.37}$$

where X_j's are as defined earlier, and E_X is a function of the h's in $D(\hat{\boldsymbol{\tau}})$ and depends on the particular factors present in the interaction.

ii) All covariances between two different interactions are identically zero, i.e.

$$Cov(\hat{\mathbf{a}}^X, \hat{\mathbf{a}}^Y) = 0 \tag{5.38}$$

if there exists at least one pair (X_j, Y_j) for which $X_j \neq Y_j$.

iii) The quadratic form

$$rvE_X \, \hat{\mathbf{a}}^{X\prime} \, \hat{\mathbf{a}}^X / (\prod_{j=1}^{n} m_j^{X_j}) \tag{5.39}$$

will be distributed as $\sigma^2 \chi^2$ with $\prod_{j=1}^{n} (m_j - 1)^{X_j}$ degrees of freedom under the hypothesis of no interaction effect.

It is clear from Theorem 5.2 that if $D(\hat{\boldsymbol{\tau}})$ takes the form (5.36), then we can find the sum of squares for various effects (main-effects and interactions) without any difficulty and all these effects will be estimated orthogonally. Now we shall identify the design patterns for which (5.36) is satisfied.

For a completely randomized or randomized block design in v treatments each having r replications, the treatment vector is estimated as

$$\hat{\boldsymbol{\tau}} = (\mathbf{I} - \mathbf{1}_v\mathbf{1}_v'/v)\ \overline{\mathbf{Y}}, \tag{5.40}$$

which has the dispersion matrix

$$D(\hat{\boldsymbol{\tau}}) = \sigma^2(\mathbf{I} - \mathbf{1}_v\mathbf{1}_v'/v)/r \tag{5.41}$$

where $\overline{\mathbf{Y}} = (\overline{Y}_1, \ldots, \overline{Y}_v)'$, $\overline{Y}_i$ being the average for the i-th treatment and σ^2 is per observation variance.

If $v = m_1 \times \ldots \times m_n$, then

$$\sum_{X(1,n)}{}' \ \mathbf{L}^X = (\mathbf{I} - \mathbf{1}_v\mathbf{1}_v')/v.$$

Taking $h_X = 1/r$ for all $\mathbf{X}$, we have

$$D(\hat{\boldsymbol{\tau}}) = \sigma^2 \sum_{X(1,n)}{}' \ h_X\mathbf{L}^X,$$

which is of the same form as in (5.36).

Similarly it can be shown that $D(\hat{\boldsymbol{\tau}})$ for a BIB design is also of the form (5.36) with r replaced by $\lambda v/k$. Thus RBD and BIB design can be used for confounded factorial experiments.

Definition 5.4: A block design $D(v, b, r, k)$ is said to possess the property (A) (or is of type PA) if its concurrence matrix $\mathbf{NN}'$ has the spectral form

$$\begin{aligned}\mathbf{NN}' &= \sum_{d=0}^{n} [\sum_{X_1+\ldots+X_n=d} \theta_{X_1\ldots X_n} \prod_{j=1}^{n} \otimes \mathbf{L}_j^{X_j}] \\ &= \sum_{X(0,n)}{}' \ \theta_X\mathbf{L}^X,\end{aligned} \tag{5.42}$$

where $\mathbf{L}_j^{X_j}$ are defined in (5.26) and $\theta_{X_1\ldots X_n}$ are the eigen values of $\mathbf{NN}'$. Clearly, $\theta_{0\ldots0} = rk$.

The form (5.42) is an alternative expression of the property (A) introduced by Kurkjian and Zelen (1963). They used the following expression to define the property (A):

$$\begin{aligned}\mathbf{NN}' &= \sum_{d=0}^{n} \{\sum_{\delta_1+\ldots+\delta_n=d} g_{\delta_1\ldots\delta_n} \prod_{=1}^{n} \otimes \mathbf{D}_j^{\delta_j}\} \\ &= \sum_{\delta(0,n)}{}' \ g_\delta \mathbf{D}^\delta,\end{aligned} \tag{5.43}$$

where $\delta_j = 0$ or 1 and

$$D_j^{\delta_j} = I_j \text{ if } \delta_j = 0$$
$$= 1_{m_j}1'_{m_j} \text{ if } \delta_j = 1.$$

Remark 5.4: The spectral form of NN' given in (5.42) is more convenient to handle as it directly gives the eigen values θ_X of NN' and the corresponding idempotent matrices L^X required for the analysis.

The M matrix of PA type design $D(v, b, r, k)$ will be

$$M = \sum_{X(0,n)}' \theta_X L^X/rk$$
$$= \sum_{X(0,n)}' \mu_X L^X, \tag{5.44}$$

where

$$\mu_X = \theta_X/rk.$$

For PA type designs, the Ω matrix for adjusted treatment mean $m\mathbf{1} + \hat{\tau}$, where m is the overall mean based on vr observations, can be written as

$$\Omega = [I + \sum_{X(1,n)}' \{\mu_X/(1-\mu_X)\} L^X]/r, \tag{5.45}$$

which on using $\sum_{X(1,n)}' L^X = I - 1_v1_v'/v$, can be written as

$$\Omega = 1_v1_v'/vr + \sum_{X(1,n)}' [1/r(1-\mu_X)] L^X. \tag{5.46}$$

Hence the dispersion matrix of $\hat{\tau}$ for a PA type design will be

$$D(\hat{\tau}) = [\sum_{X(1,n)}' \{1/r(1-\mu_X)\} L^X]\sigma^2. \tag{5.47}$$

Comparing (5.36) and (5.47) we get

$$h_X = 1/r(1-\mu_X).$$

The readers may take note of the following results on PA type designs:

(i) the estimate $\hat{\tau} = \Omega Q$ will be

$$\hat{\tau} = [\sum_{X(1,n)}' \{1/r(1-\mu_X)\} L^X] Q$$
$$= [\sum_{X(1,n)}' \{1/rE_X\} L^X] Q$$

because $1'Q = 0$, where $E_X = 1 - \mu_X$.

(ii) the interaction vector a^X may be estimated from the adjusted treatment total Q as follows:

$$\hat{a}^X = C^X\hat{\tau}$$

$$= [\sum_{Y(1,n)}' \{1/rE_Y\} \mathbf{C}^X\mathbf{L}^Y] \mathbf{Q}$$

$$= \sum_{d=1}^{n} [\sum_{Y_1+\ldots+Y_n=d} \{1/rE_{Y_1\ldots Y_n}\}\} \prod_{j=1}^{n} \otimes \mathbf{C}_j^{X_j}\mathbf{L}_j^{Y_j}]$$

$$= [\prod_{j=1}^{n} \otimes \mathbf{C}_j^{X_j}] \mathbf{Q}/rE_{X_1\ldots X_n}, \tag{5.48}$$

because

$$\mathbf{C}_j^{X_j}\mathbf{L}_j^{Y_j} = \mathbf{C}_j^{X_j} \text{ if } X_j = Y_j$$

$$= 0 \text{ otherwise.}$$

The quantity $E_{X_1\ldots X_n}$ is the efficiency factor associated with the generalized interaction $\mathbf{a}_1^{X_1} \times \ldots \times \mathbf{a}_n^{X_n}$. Although there are $2^n - 1$ such efficiency factors many of them may coincide.

(iii) $D(\hat{\mathbf{a}}^X) = \mathbf{C}^X D(\hat{\tau}) \mathbf{C}^{X'}$

$$= [\sum_{Y(1,n)}' \{1/rE_Y\} \mathbf{C}^X\mathbf{L}^X\mathbf{C}^{X'}] \sigma^2$$

$$= \sigma^2 [\prod_{j=1}^{n} \otimes \mathbf{C}_j^{X_j}\mathbf{C}_j^{X_j'}]/rE_{X_1\ldots X_n}$$

$$= \sigma^2 [\prod_{j=1}^{n} m_j^{X_j}\mathbf{C}_{j_1} \otimes \mathbf{C}_{j_2} \otimes \ldots \otimes \mathbf{C}_{j_p}]/rvE_{X_1\ldots X_n} \tag{5.49}$$

because

$$\mathbf{C}_j^{X_j}\mathbf{C}_j^{X_j'} = \mathbf{C}_j \text{ if } X_j = 1$$

$$= 1/m_j \text{ if } X_j = 0.$$

(iv) In a similar fashion it can be shown that covariance matrix between two different interaction vectors is zero, i.e.

$$Cov\ (\hat{\mathbf{a}}^X, \hat{\mathbf{a}}^Y) = \mathbf{0}, \tag{5.50}$$

if there exists at least one pair (X_j, Y_j) for which $X_j \neq Y_j$.

(v) Note that by (5.49)

$$[D(\hat{\mathbf{a}}^X)]^2 = \{\prod_{j=1}^{n} m_j^{X_j}/rvE_{X_1\ldots X_n}\} D(\hat{\mathbf{a}}^X) \sigma^2$$

$$= \phi_X D(\hat{\mathbf{a}}^X) \sigma^2 \tag{5.51}$$

where

$$\phi_X = \prod_{j=1}^{n} m_j^{X_j}/rvE_{X_1\ldots X_n}.$$

The condition (5.51) together with the assumption of normality implies that the quadratic form

$$\hat{\mathbf{a}}^{X'}\hat{\mathbf{a}}^{X}/\phi_X = [rvE_X\hat{\mathbf{a}}^{X'}\hat{\mathbf{a}}^{X}]/\prod_{j=1}^{n} m_j^{X_j} \tag{5.52}$$

has a $\sigma^2\chi^2$ distribution with $\prod_{j=1}^{n}(m_j-1)^{X_j}$ degrees of freedom under the assumption of no interaction effect.

We therefore find that PA type designs are factorially balanced. Kshirsagar (1966) has proved the converse of this result viz., that the factorially balanced designs are PA type and are infact EGD/2^n-1. Kshirsagar's work is also significant as it provides a reciprocal relation between the structural forms (5.42) and (5.43) of $\mathbf{NN}'$.

The parameters of a factorially balanced design can be written as

$$v = m_1 \times \ldots \times m_n,\ b,\ r,\ k,\ \mu_{X_1 \ldots X_n},\ X_j = 0 \text{ or } 1,$$

such that not all X_j's are zero simultaneously. The adjusted treatment sum of squares, $\mathbf{Q}'\mathbf{\Omega Q}$, for factorially balanced designs gets simplified to

$$\sum_{X(1,\,n)}' \{1/r(1-\mu_X)\}\,\mathbf{Q}'\mathbf{L}^X\mathbf{Q} = \sum_{X(1,\,n)}' (\text{sum of squares for the effect } F_1^{X_1}\ldots F_n^{X_n}),$$

where the sum of squares for the interaction $F_1^{X_1}\ldots F_n^{X_n}$ with $(m_1-1)^{X_1}\ldots(m_n-1)^{X_n}$ degrees of freedom is $\{1/r(1-\mu_X)\}\mathbf{Q}'\mathbf{L}^X\mathbf{Q}$, which can alternatively be written in the form (5.34) with $h_X = 1/r(1-\mu_X)$.

From the above discussions it is clear that if the values of $\mu_{X_1 \ldots X_n}$ are known in any factorially balanced design, then the remaining analysis is simple and straightforward. For the sake of futher clarity, we examine some of the well known Kronecker product designs as factorials and give simplified expressions for their analysis.

5.4.1 Balanced incomplete block designs

Consider a BIB design with parameters v, b, r, k, λ. We have already seen in Chapter 4 that the concurrence matrix of the BIB design can be written as

$$\begin{aligned}\mathbf{NN}' &= (r-\lambda)[\mathbf{I}_v - \mathbf{11}'/v] + rk\mathbf{11}'/v\\ &= (r-\lambda)\mathbf{L}_1 + rk\mathbf{L}_0,\end{aligned}$$

where $\mathbf{L}_0 = \mathbf{11}'/v$, $\mathbf{L}_1 = \mathbf{I}_v - \mathbf{11}'/v$.

If $v = \prod_{j=1}^{n} m_j$, then the BIB design can be used as an $m_1 \times \ldots \times m_n$ factorial experiment. If $\mathbf{L}_j^{X_j}$, $j = 1, \ldots, n$ are as defined in (5.26), then

$$\sum_{X(1,\,n)}' \prod_{j=1}^{n} \otimes\, \mathbf{L}_j^{X_j} = (\mathbf{I}_v - \mathbf{11}'/v)$$

and the concurrence matrix can be written as

$$NN' = rk\mathbf{11}'/v + (r-\lambda) \sum_{X(1,\,n)}' L^X$$
$$= \sum_{X(0,\,n)}' \theta_X L^X,$$

where $\theta_{X_1 \ldots X_n} = rk$ if $X_j = 0$, $j = 1, \ldots, n$,
$= r - \lambda$ otherwise.

This leads to the following theorem:

Theorem 5.3: The existence of a BIB design with parameters

$$v = \prod_{j=1}^{n} m_j,\ b,\ r,\ k,\ \lambda$$

implies the existence of a BFE with parameters

$$v = \prod_{j=1}^{n} m_j,\ b,\ r,\ k,\quad \mu_{X_1 \ldots X_n} = (r-\lambda)/rk.$$

As pointed out earlier this class of designs has little utility as a factorial experiment because all its effects are confounded with the blocks to the same extent.

5.4.2 Kronecker product PBIB designs

This class of designs has been discussed in Chapter 4 where it was shown that NN′ matrix of such designs equals $\Sigma' \theta_X L^X$. It therefore follows that these designs are balanced factorials. As stated earlier, this class includes a wide variety of PBIB designs which can be derived as reduced forms of EGD/$2^n - 1$ type PBIB designs. PBIB designs falling into this category are group divisible, latin square (L_2), rectangular, cubic and hypercubic. We now examine this class in further details.

5.4.2.1: Group divisible designs

It is easy to prove the following theorem:

Theorem 5.4: The existence of GD design with parameters

$$v = m_1 m_2,\ b,\ r,\ k,\ \lambda_1,\ \lambda_2,\ n_1 = (m_2 - 1),\ n_2 = m_2(m_1 - 1)$$

implies the existence of a BFE with parameters

$$v = m_1 m_2,\ b,\ r,\ k,\ \mu_{10} = (rk - v\lambda_2)/rk,\ \ \mu_{01} = \mu_{11} = (r - \lambda_1)/rk.$$

The factorial designs constructed from semi-regular and singular GD designs have special significance as in case of semi-regular GD the main effect F_1 and in case of singular the main effect F_2 and interaction F_1F_2 are unconfounded with the blocks and we have the following corollaries:

Corollary 5.1: The existence of semi-regular GD design implies the existence of a BFE with parameters

$$v = m_1 m_2, b, r, k, \mu_{10} = 0, \mu_{01} = \mu_{11} = (r - \lambda_1)/rk.$$

Corollary 5.2: The existence of singular GD design implies the existence of BFE with parameters

$$v = m_1 m_2, b, r, k, \ \mu_{10} = rk - v\lambda_2, \mu_{01} = \mu_{11} = 0.$$

The GD design can be utilized for multiple factor factorials replacing the levels of F_1 factor by the factorial combinations of factors B_1, ..., B_{n_1} and the levels of F_2 factor by factorial combinations of C_1, ..., C_{n_2} factors, where the levels of these factors are t_1, ..., t_{n_1}, t_{n_1+1}, ...,$t_{n_1+n_2}$, respectively, such that

$$\sum_{j=1}^{n_1} t_j = m_1 \text{ and } \sum_{j=1}^{n_2} t_{n_1+j} = m_2.$$

A typical interaction will be denoted by

$$b_1^{X_1} \otimes \cdots \otimes b_{n_1}^{X_{n_1}} \otimes c_1^{X_{n_1+1}} \otimes \cdots \otimes c_{n_2}^{X_n},$$

where $n = n_1 + n_2$. If $\sum_{j=n_1+1}^{n} X_j = 0$, the interaction consists of only B-factors and if $\sum_{j=1}^{n_1} X_j = 0$, it consists of C-factors only.

It is easy to show that,

$$\sum_{X(1,\, n_1)}' \mathbf{L}^X = \mathbf{I}_{n_1} - \mathbf{1}_{n_1}\mathbf{1}'_{n_1}/n_1$$

and

$$\sum_{X(1,\, n_2)}' \mathbf{L}^X = \mathbf{I}_{n_2} - \mathbf{1}_{n_2}\mathbf{1}'_{n_2}/n_2$$

and

$$\mathbf{NN}' = \sum_{X(0,\, n)}' \theta_X \mathbf{L}^X,$$

where

$$\begin{aligned} \theta_{X_1 \ldots X_{n_1} X_{n_1+1} \ldots X_{n_1+n_2}} &= rk && \text{if } X_j = 0 \\ &= rk - v\lambda_2 && \text{if } \sum_{j=n_1+1}^{n} X_j = 0 \\ &= r - \lambda_1 && \text{otherwise.} \end{aligned}$$

We therefore, have the following theorem:

Theorem 5.5: The existence of a GD PBIB design with parameters

$$v = m_1 m_2, b, r, k, \lambda_1, \lambda_2, n_1 = (m_2 - 1), n_2 = m_2 (m_1 - 1)$$

with $m_1 = \prod_{j=1}^{n_1} t_j$ and $m_2 = \prod_{j=1}^{n_2} t_{n_1+j}$ implies the existence of a BFE with parameters

$$v = t_1 \times \ldots \times t_n,\ b,\ r,\ k,\ \mu_{x_1x_2\ldots x_m x_{n_1+1}\ldots x_n} = (rk - v\lambda_2)/rk$$

if $\sum_{j=n_1+1}^{n} X_j = 0$ and $(r - \lambda_1)/rk$ otherwise.

Consider the singular GD PBIB design listed as S 106 in Clatworthy (1973) with parameters

$$v = 20,\ b = 6,\ r = 3,\ k = 10,\ m_1 = 4,\ m_2 = 5,\ \lambda_1 = 3,\ \lambda_2 = 1.$$

Then we can construct the BFE 5×4 and 5×2^2 in 3 replications when only F_1 effect is confounded with relative loss of information 1/3.

5.4.2.2 L_2 PBIB designs

L_2 PBIB design can also be used to provide a two-factor symmetrical factorial or multifactor factorial experiment $t_1 \times \ldots \times t_n$ if

$$m = \prod_{j=1}^{n_1} t_j = \prod_{j=1}^{n_2} t_{n_1+j}.$$

We can prove the following theorem on the similar lines of theorems 5.4 and 5.5.

Theorem 5.6: The existence of L_2 PBIB design with parameters

$$v = m^2,\ b,\ r,\ k,\ \lambda_1,\ \lambda_2,\ n_1 = 2\ (v - 1) \text{ and } n_2 = (v - 1)^2$$

implies the existence of BFE with parameters

$$v = m^2,\ b,\ r,\ k,\ \mu_{01} = \mu_{10} = [r - \lambda_1 + (m-1)(\lambda_1 - \lambda_2)]/rk$$

and $\mu_{11} = (r - 2\lambda_1 + \lambda_2)/rk$

and a BFE with parameters

$$v = t_1 \times \ldots \times t_n,\ b,\ r,\ k,\ \mu_{x_1\ldots x_{n_1} x_{n_1+1}\ldots x_n}$$

$$= \{r - \lambda_1 + (m_1 - 1)(\lambda_1 - \lambda_2)\}/rk \text{ if } \sum_{j=1}^{n_1} X_j = 0 \text{ or } \sum_{j=n_1+1}^{n} X_j = 0$$

and $(r - 2\lambda_1 + \lambda_2)/rk$ otherwise.

5.4.2.3 Rectangular PBIB designs

Three associate class rectangular PBIB designs also possess the property (A) and therefore can be used as two-factor factorial experiments. The following theorem can be proved easily:

Theorem 5.7: The existence of a rectangular PBIB design with parameters

$$v = m_1m_2,\ b,\ r,\ k,\ \lambda_1,\ \lambda_2,\ \lambda_3,\ n_1 = m_2 - 1,\ n_2 = m_1 - 1 \text{ and}$$

$$n_3 = (m_1 - 1)\ (m_2 - 1)$$

implies the existence of a BFE with parameters

$$v = m_1 m_2,\ b,\ r,\ k,\ \mu_{01} = r - \lambda_1 + (m_1 - 1)(\lambda_2 - \lambda_3),$$

$$\mu_{10} = r - \lambda_2 + (m_2 - 1)(\lambda_1 - \lambda_3),\ \mu_{11} = r - \lambda_1 - \lambda_2 + \lambda_3.$$

We can utilize a rectangular PBIB design to construct a multifactor $t_1 \times \ldots \times t_{n_1} \times t_{n_1+1} \times \ldots \times t_n$ factorial by assigning m_1 factorial combinations among the B factors to m_1 rows and m_2 factorial combinations to m_2 columns, where $m_1 = \prod_{j=1}^{n_1} t_j$ and $m_2 = \prod_{j=n_1+1}^{n} t_j$.

It is easy to prove the following theorem:

Theorem 5.8: The existence of a rectangular PBIB design with parameters

$$v = m_1 m_2,\ b,\ r,\ k,\ \lambda_1,\ \lambda_2,\ \lambda_3,\ n_1 = m_2 - 1,\ n_2 = m_1 - 1 \text{ and}$$
$$n_3 = (m_1 - 1)(m_2 - 1)$$

implies the existence of a BFE with parameters

$$v = t_1 \times \ldots \times t_n,\ b,\ r,\ k,$$

$$\mu_{X_1 \ldots X_{n_1} \ldots X_n} = [r - \lambda_1 + (m_1 - 1)(\lambda_2 - \lambda_3)]/rk \quad \text{if } \sum_{j=1}^{n_1} X_j = 0$$

$$= r\,[r - \lambda_2 + (m_2 - 1)(\lambda_1 - \lambda_3)]/rk \quad \text{if } \sum_{j=n_1+1}^{n} X_j = 0,$$

$$= r - \lambda_1 - \lambda_2 + \lambda_3 \text{ otherwise.}$$

5.4.2.4 EGD and other designs

It can be easily shown that the other Kronecker Product PBIB designs also possess property (A) and hence can be utilized as factorial experiments. We can easily prove the following theorem:

Theorem 5.9: The existence of EGD/2^n-1 PBIB design with parameters

$$v = \prod_{j=1}^{n} m_j,\ b,\ r,\ k,\ n_{X_1 \ldots X_n},\ \lambda_{X_1 \ldots X_n},$$

$$n_{X_1 \ldots X_n} = \prod_{j=1}^{n} (n_j - 1)^{X_j},$$

implies the existence of a BFE with parameters

$$v = m_1 \times \ldots \times m_n,\ b,\ r,\ k,\ \mu_{X_1 \ldots X_n} = \theta_{X_1 \ldots X_n}/rk$$

where $\theta_{X_1 \ldots X_n}$ are the eigen values of NN' matrix.

Cubic designs can be used to construct three factor symmetrical factorial experiments where all main effects are confounded with blocks to same extent and all two factor interactions are also confounded to same extent.

If we denote $\mu_{X_1X_2X_3} = \mu_i$ if $\Sigma X_j = i$ then it is easy to prove the following theorem:

Theorem 5.10: The existence of cubic designs with parameters $v = m^3, b, r, k, \lambda_1, \lambda_2, \lambda_3, n_1 = 3(m-1), n_2 = 3(m-1)^2, n_3 = (m-1)^3$ implies the existence of three factor factorial experiments with parameters

$$v = m^3, b, r, k, \mu_1 = \mu_{100} = \mu_{010} = \mu_{001} = \theta_1/rk,$$

$$\mu_2 = \mu_{110} = \mu_{101} = \mu_{011} = \theta_2/rk \text{ and } \mu_3 = \mu_{111} = \theta_3/rk,$$

where $\theta_i (i = 1, 2, 3)$ are the eigen values of **NN**$'$ matrix.

Similarly, hypercubic designs can be used to construct m^n factorial experiments. The generalized right angular PBIB designs with parameters

$$v = m_1m_2m_3, b, r, k, \lambda_1, \lambda_2, \lambda_3, \lambda_4, n_1 = (m_3 - 1),$$

$$n_2 = m_3(m_2 - 1), n_3 = m_3(m_1 - 1) \text{ and } n_4 = m_3(m_1 - 1)(m_2 - 1)$$

provide three factor $m_1 \times m_2 \times m_3$ factorial experiments where main effects F_1, F_2 and interaction F_1F_2 are estimated with losses θ_3/rk, θ_2/rk and θ_4/rk, respectively, and remaining effects with relative losses θ_1/rk, where θ_i's are the eigen values of **NN**$'$.

The above designs can also be used for multi-factor factorial experiments by replacing the levels of first factor F_1 by the factorial combinations of factors $B_1, \ldots, B_{n_1}$ with levels $t_1, \ldots, t_{n_1}$, the levels of the second factor F_2 by the level combinations $t_{n_1+1}, \ldots, t_{n_1+n_2}$ of factors $C_1, \ldots, C_{n_2}$ and so on where $m_1 = \prod_{j=1}^{n_1} t_j$, $m_2 = \prod_{j=n_1+1}^{n} t_j$ and so on.

Before we conclude this Section, we consider again the four replicate design of Das (1960a) for 3×2^2 experiment. It was pointed out that the design formed by first or last two replicates has factorial structure. These designs are not factorially balanced, but the design with all the four replicates is factorially balanced.

5.5. SOME OTHER RELATED RESULTS

Puri and Nigam (1976) introduced the structural property (A^*) as a generalization of property (A) to include the designs with varying replications and varying block sizes.

Consider a block design D with parameters $v = \prod_{j=1}^{n} m_j$, b, **r**, **k**. Let us further assume that **r** can be expressed as

$$\mathbf{r} = r(\boldsymbol{\alpha}_{m_1} \otimes \cdots \otimes \boldsymbol{\alpha}_{m_n}),$$

where $\boldsymbol{\alpha}_{m_j}$, an $m_j \times 1$ vector, represents the relative replications for the j-th factor and r is some constant. In case of equi-replicate designs $\boldsymbol{\alpha}_{m_j} = \mathbf{1}_{m_j}$

and r denotes the replication for any treatment. It may be remarked here that this choice of $\mathbf{r}$ in the context of factorial experiment is not unrealistic. The structural property (A^*) is defined as follows:

Definition 5.5: A block design $D(v, b, \mathbf{r}, \mathbf{k})$ is said to possess the property (A^*) if its $\mathbf{M}$ matrix has the spectral form

$$\mathbf{M} = \sum_{d=0}^{n} \{ \sum_{X_1+\ldots+X_n=d} \mu_{X_1\ldots X_n} \prod_{j=1}^{n} \otimes \mathbf{L}_j^{X_j} \}$$

$$= \sum_{X(0,n)}' \mu_X \mathbf{L}^X,$$

where

$$\mathbf{L}_j^{X_j} = \mathbf{I}_{m_j} - \mathbf{1}_{m_j}\alpha'_{m_j}/m_j \quad \text{if} \quad X_j = 1,$$

$$= \mathbf{1}_{m_j}\alpha'_{m_j}/m_j \quad \text{otherwise,}$$

and
$$\alpha'_{m_j}\mathbf{1}_{m_j} = m_j.$$

For the PA*-type designs the $\mathbf{\Omega}$ matrix will be

$$\mathbf{\Omega} = \mathbf{1}\mathbf{1}'/n + \sum_{X(1,n)}' [1/(1-\mu_X)]\, \mathbf{L}^X\mathbf{R}^{-1}.$$

If $\mathbf{Q}$ denotes the vector of adjusted treatment totals, then the adjusted treatment sum of squares will be

$$\sum_{X(1,n)}' \mathbf{Q}'\mathbf{L}^X\mathbf{R}^{-1}\mathbf{Q}/(1-\mu_X),$$

and sum of squares due to any interaction $F_1^{X_1}\ldots F_n^{X_n}$ will be

$$\mathbf{Q}'\mathbf{L}^X\mathbf{R}^{-1}\mathbf{Q}/(1-\mu_X).$$

If $D_j(j = 1, \ldots, n)$ are n efficiency balanced designs with parameters $v=m_j, b_j, \mathbf{r}_j, \mathbf{k}_j, \mu_j$ and incidence matrices $\mathbf{N}_j$, then Puri and Nigam (1977) have shown that

$$\mathbf{N} = \prod_{j=1}^{n} \otimes \mathbf{N}_j$$

is a PEB design with atmost 2^n-1 efficiency classes with parameters

$$v = \prod_{j=1}^{n} m_j,\ b,\ \mathbf{r},\ \mathbf{k},\ \mu_{X_1\ldots X_n},$$

where

$$\mu_{X_1\ldots X_n} = \prod_{j=1}^{n} \mu_j^{X_j},$$

are the eigen values of $\mathbf{M}$ matrix with multiplicities $\prod_{j=1}^{n} (m_j-1)^{X_j}$. These designs are called EGD/2^n-1 PEB designs. For EGD/2^n-1 PEB designs the $\mathbf{M}$ matrix will have the spectral decomposition

$$\mathbf{M} = \sum_{X(0,n)}' \mu_X \mathbf{L}^X,$$

where

$$\mathbf{L}_j^{X_j} = \mathbf{I}_j - \mathbf{1}_j \mathbf{r}_j'/n_j \quad \text{if} \quad X_j = 1$$
$$= \mathbf{1}_j \mathbf{r}_j'/n_j \quad \text{if} \quad X_j = 0$$

and $n_j = \mathbf{r}_j'\mathbf{1}_j$. Here $\mathbf{r} = \mathbf{r}_1 \otimes \ldots \otimes \mathbf{r}_n$. Hence, EGD/$2^n - 1$ type PEB designs are PA* type designs. The designs obtained from Kronecker Product designs after merging the treatments can also be shown to be PA* type designs. For some details, the interested reader may refer to Puri and Nigam (1976, 1978).

Besides the work of Puri and Nigam, other interesting findings are from Mukherjee (1979, 1980, 1981) and Gupta (1983). Mukherjee improved the results of John and Smith (1972) and John, Cotter and Smith (1973) and obtained a necessary and sufficient condition on **NN′** for the design to have factorial structure. He defined a structure (K) which means that **NN′** can be expressed as a linear combination of Kronecker products of square matrices with same row and column totals. He showed that a design has factorial structure if and only if **NN′** has structure (K). Gupta (1983) extended the scope of property (A) of Kurkjian and Zelen by expressing **A** matrix in a form similar to (5.42). The interested readers may refer to these papers for details.

6. Cyclic Designs

6.1 INTRODUCTION

In the earlier Chapters we have discussed balanced and partially balanced designs and designs with factorial structure. We now discuss another very useful class of designs known as cyclic designs, which are of utility both as block designs and as designs with factorial structure.

Cyclic designs are incomplete block designs obtained through cyclic development of one or more initial blocks. Their flexibility, ease of representation and experimentation make them worthy of attention although their analysis is slightly more involved than that of more commonly used balanced and partially balanced designs discussed thus far. Cyclic designs as a class were introduced by Kempthorne (1953) and were extensively studied by David (1963, 1965), Wolock (1964), David and Wolock (1965) and John (1966). In a later work, John (1969) studied the relationship between cyclic and PBIB designs and showed that cyclic designs in v treatments can be PBIB with atmost $v/2$ associate classes. These designs also resemble closely with circular designs of Das (1960b). In view of the tremendous potential of cyclic designs, John, Wolock and David (1972) tabulated these designs in the range $6 \leqslant v \leqslant 60$, $2 \leqslant r \leqslant k \leqslant 10$ and gave the average efficiency factors of different designs.

6.1.1 Generating methods

A cyclic design can be denoted by the triplet of parameters (v, k, r). Let the treatments be labelled as 0, 1,. ., $v-1$. Then if the i-th block contains the treatments a_0, a_1, .., a_{k-1} then the $(i+j)$-th block will have the treatments a_0+j, a_1+j,..., $a_{k-1}+j$, where $a_h = 0, 1, \ldots, v-1$; $h = 0, 1, \ldots, k-1$ and $a_h + j$ is reduced to mod (v).

To illustrate the idea, consider the seven blocks generated by developing cyclically the initial block (0, 1, 2) mod (7). The seven blocks are (0, 1, 2), (1, 2, 3), (2, 3, 4), (3, 4, 5), (4, 5, 6), (5, 6, 0), (6, 0, 1). This design is partially balanced with three efficiency factors. If the initial block is taken as (0, 1, 3) mod (7), then the generated blocks are (0, 1, 3), (1, 2, 4), (2, 3, 5), (3, 4, 6), (4, 5, 0), (5, 6, 1), (6, 0, 2). The generated design is symmetrical BIB with parameters $v = b = 7$, $r = k = 3$, $\lambda = 1$. This flexibility of generating a balanced or a partially balanced design is the main strength of this class of designs.

Cyclic designs can also be obtained by combining more than one cyclic set for a given v. For instance, the two sets (0, 1, 2) and (0, 1, 3) can be combined to give a partially balanced design for 7 treatments in 14 blocks. Again, if we combine the cyclic sets (0, 1, 2), (0, 1, 3), (0, 1, 4), (0, 1, 5) and (0, 2, 4) we get the set of all the $\binom{7}{3}$ distinct blocks which form the irreducible BIB design with parameters $v = 7$, $b = 35$, $r = 15$, $k = 3$, $\lambda = 5$. It is also possible to construct BIB designs with repeated blocks. For instance, we can repeat five times the set of 7 blocks obtained by developing cyclically the initial block (0, 1, 3) mod (7). The derived arrangement will also be a BIB design $v = 7$, $b = 35$, $r = 15$, $k = 3$, $\lambda = 5$ in which only 7 blocks will be distinct. Such BIB designs with repeated blocks have been studied by Wynn (1977), Foody and Hedayat (1977) and John (1978). In a recent work, Nigam, Kumar and Gupta (1984) have found these designs useful in problems relating to inclusion probability proportional to size sampling.

It is realized that in cyclic designs derived from a single cyclic set the number of replications for each treatment and the number of blocks are equal to block size and the number of treatments respectively. If the design is obtained by combining several cyclic sets then these numbers are multiples of the number of cyclic sets taken. The number of treatments and the block sizes, however, remain the same. Since the replication of each treatment and the number of blocks become very large whenever v is large, the adoption of cyclic designs becomes unmanageable from experimenter's point of view. This is a serious limitation of cyclic designs. To overcome this problem the idea of fractional sets was introduced which we describe in the sequel.

6.2 FRACTIONAL SETS

In a single cyclic set (v, k, r) the value of r equals k and in the case of p (> 1)-cyclic sets $r = pk$. The use of fractional sets helps in reducing the value of r to less than k. If v and k have a common divisor d, then a fractional set $(v, k, k/d)$ exists. We illustrate this point by constructing the partially balanced design with parameters $v = 14$, $b = 7$, $k = 6$, $r = 3$. The design can be obtained by using two BIB designs in 7 treatments obtained through single cyclic sets (0, 1, 3) and (7, 8, 10). The treatments in the first design are labelled as 0, 1,..., 6 and the contents of the initial block are (0, 1, 3) while the treatments of the second design are labelled as 7, 8,..., 13 with the initial block as (7, 8, 10). We can now obtain 7 blocks each of size 6 by developing (0, 1, 3) and (7, 8, 10) using cycles of length 7 and cycling 0 to 6 and 7 to 13 respectively in the cyclic set (0, 1, 3; 7, 8, 10). The complete solution of the design with $r = 3$ is listed below:

0	1	3	7	8	10
1	2	4	8	9	11
2	3	5	9	10	12
3	4	6	10	11	13
4	5	0	11	12	7
5	6	1	12	13	8
6	0	2	13	7	9

The design is in fact singular **GD** with the parameters $v = 14, b = 7, k = 6, r = 3, \lambda_1 = 3, \lambda_2 = 1, m = 7, n = 2$.

Remark 6.1: The idea of fractional sets just described once again does not appear to be satisfactory from experimenter's viewpoint because the block size becomes large. In fact if there are p cyclic sets, then the block size of the resulting design is pk. An alternate way could be to obtain a design $v = b = 14, r = k = 3$ in the following manner. The blocks of this design are (0, 1, 3), (1, 2, 4), (2, 3, 5), (3, 4, 6), (4, 5, 0), (5, 6, 1), (6, 0, 2), (7, 8, 10), (8, 9, 11), (9, 10, 12), (10, 11, 13), (11, 12, 7), (12, 13, 8), (13, 7, 9). Obviously this design is disconnected. Another arrangement could be (0, 1, 3, 7), (1, 2, 4, 8), (2, 3, 5, 9), (3, 4, 6, 10), (4, 5, 0, 11), (5, 6, 1, 12), (6, 0, 2, 13), (7, 8, 10, 0), (8, 9, 11, 1), (9, 10, 12, 2), (10, 11, 13, 3), (11, 12, 7, 4), (12, 13, 8, 5), (13, 7, 9, 6) obtained by developing two initial sets (0, 1, 3, 7) and (7, 8, 10, 0), the cycles of development being the same as described above in the Example. The parameters of the resulting design are $v = b = 14$, $r = k = 4$. A detailed study on the resulting designs obtained by using this method, however, needs attention.

6.3 RESOLVABLE CYCLIC DESIGNS

By the very method of developing a cyclic set (v, k, r) mod (v) it is immediate that the design generated through p cyclic sets is r-resolvable. A further source of variation can therefore be removed by considering each of the r-resolvable sets as a natural stage of experimentation. As an example consider again the BIB design $v = 7$, $b = 35$, $k = 3$, $r = 15$, $\lambda = 5$ obtained through the five cyclic sets (0, 1, 2), (0, 1, 3), (0, 1, 4), (0, 1, 5) and (0, 2, 4). The design is 3-resolvable with each of the 5 cyclic sets forming a 3-resolvable cyclic set. In many situations it may be possible to rearrange the blocks to construct another design which is 1- or more resolvable. Consider, for example, the 3-resolvable design in two sets of six blocks each obtained by cyclically developing the initial blocks (0, 1, 3) and (2, 4, 5) mod (6). From this design a 1-resolvable design in six sets of two blocks each can be obtained by combining the complementary blocks from the two sets to form a single set. Thus, (0, 1, 3) and (2, 4, 5) will form one resolvable set, (1, 2, 4) and (3, 5, 0) will form the second resolvable set, and so on, and (5, 0, 2) and (1, 3, 4) will form the sixth resolvable set.

For a cyclic design through the single cyclic set (v, k, r) to be resolvable a necessary condition is that v/k be an integer. For example if we consider the initial set (1, 4) mod (6), then the generated blocks are (1, 4), (2, 5), (3, 0), (4, 1), (5, 2), (0, 3). These can be divided into two resolvable sets of three blocks each.

6.4 ANALYSIS OF CYCLIC DESIGNS

We have seen in Chapter 2 that if **N** is the incidence matrix of any block design, then the intrablock estimates of the treatment effects, $\boldsymbol{\tau}$ can be obtained from the standard normal equations $\mathbf{A}\,\hat{\boldsymbol{\tau}} = \mathbf{Q}$, where for an equi-replicate and proper design $\mathbf{A} = r\,\mathbf{I} - (1/k)\,\mathbf{NN}'$. A solution to the system of normal equations is $\hat{\boldsymbol{\tau}} = \mathbf{A}^{-}\mathbf{Q}$, where $\mathbf{A}^{-}$ is any generalized inverse of **A** and **Q** denotes the vector of adjusted treatment totals. For cyclic designs the matrices $\mathbf{NN}'$, **A**, $\boldsymbol{\Omega}^{-1}$ and $\boldsymbol{\Omega}$ are all circulant. As these designs are equi-replicate and proper, it is sufficient to work with matrix **A**. Since **A** is a circulant matrix, $\mathbf{A}^{-}$ will also be a circulant matrix and can be obtained by a method described by Kempthorne (1953). The elements of the first row of $\mathbf{A}^{-}$ are

$$a^{11} = \frac{1}{v} \sum_{h=2}^{v} 1/\theta_h$$

$$a^{1j} = \frac{1}{v} \sum_{h=2}^{t} \frac{\cos (h-1)(j-1)\,\phi}{\theta_h},\ j = 2,\ldots, v$$

where $\phi = 2\pi/v$ and θ_j is an eigen value defined by

$$\theta_j = a_1 + \sum_{h=2}^{v} a_h \cos (h-1)(j-1)\,\phi.$$

The eigen values θ_j are clearly symmetric with θ_{v-j+2}. The a_i $(i = 1,..., v)$ values are determined by the association scheme of the design, viz.,

$$a_1 = r\,(k-1)/k,\ a_2 = a_v = -\lambda_{12}/k,\ a_3 = a_{v-1} = -\lambda_{13}/k,$$
$$\ldots\ a_{(v+1)/2} = a_{(v+3)/2} = -\lambda_{1(v+1)/2}/k, \text{ if } v \text{ is odd, cr}$$
$$\ldots\ a_{(v+2)/2} = -\lambda_{1(v+2)/2}/k, \text{ if } v \text{ is even,}$$

where λ_{ij} is the number of blocks in which treatments i and j appear together. The adjusted sum of squares due to treatments is

$$\hat{\boldsymbol{\tau}}'\mathbf{Q} = \sum_{i=1}^{v} \hat{\tau}_i\, Q_i$$

where $$\mathbf{Q} = (Q_1,..., Q_v)'.$$

6.5 APPLICATIONS OF CYCLIC DESIGNS

Among the class of binary incomplete block designs, the BIB and PBIB designs have been studied extensively. Apart from the BIB designs PBIB designs with two associate classes have also been found to be useful in practice because of their structural properties and simple analysis. But these designs are not available for many parametric combinations (v, k, r) of interest. This makes the cyclic designs useful as these designs supplement tremendously the already existing class of BIB and PBIB designs. It is seen that cyclic designs admit a straightforward and explicit analysis because their $\boldsymbol{\Omega}$ matrix is circulant.

Cyclic designs have been found to be extremely useful in diallel experiments. If there are v parents then there are $\binom{v}{2}$ crosses among them. Experiments aimed at estimating the 'general' and 'specific' combining abilities of the v lines (parents) are called diallel crosses or diallel experiments. When v is large, such an experiment will result in excessively large number of crosses. In such situations sampling of the crosses is resorted to. Experiments requiring only a subset of $\binom{v}{2}$ possible crosses are known as partial diallel cross experiments. The idea of crossing two inbred lines can easily be extended to three lines, giving rise to triallel experiments. A good amount of literature is available on diallel and triallel crosses, partial as well as complete. Notable contributions are by Curnow (1963), Hinkelmann (1960, 1965, 1966), Kempthorne and Curnow (1961), Fyfe and Gilbert (1963), and Das and Sivaram (1968). Restricting our discussion to the use of cyclic designs, it may be stated that these were used as 'circulant samples' for partial diallel crosses by Kempthorne and Curnow (1961) and Fyfe and Gilbert (1963).

Cyclic designs are also used for paired comparisons. One of the applications of paired comparisons is in taste testing experiments where a judge may be asked to give his preference for one flavour or the other in each of the $\binom{v}{2}$ pairwise comparisons (flavours) involving v juices. If more than one judge (say, n) is available, then as many replicates are considered. When v is large it may be virtually impossible for a judge to differentiate between $\binom{v}{2}$ pairwise comparisons. Hence, there is a need to consider a subset of $\binom{v}{2}$ comparisons. Mckeon (1960) considered this problem and found that cyclic designs are much useful for the purpose. Another interesting application of cyclic designs has recently been suggested by Nigam, Kumar and Gupta (1984). They have found these designs useful in inclusion probability proportional to size sampling. Another usefulness of symmetrical BIB designs obtained through initial block solutions has been demonstrated by Gupta, Nigam and Kumar (1982) in obtaining a family of inclusion probability proportional to size sampling schemes.

There are several other interesting applications of cyclic designs. These have been used as 'Tournament designs' (David, 1967) and also in caliberation experiments (Cameron and Speckman, 1972). Interested readers may refer to these for further reading.

6.6 GENERALIZED CYCLIC DESIGNS

Generalized cyclic designs (GC/n) have been studied extensively by John (1973) as a generalization of cyclic designs (GC/1). These designs are extremely useful for partial confounding in factorial experiments.

Consider an incomplete block design $D(v, b, r, k)$, where v can be decomposed as $v = m_1 \ldots m_n$. Let us label the $v = \prod_{j=1}^{n} m_j$ treatments or treatment combinations in case of $m_1 \times \ldots \times m_n$ factorial experiment by the n-tuple

$$i = (i_1, \ldots, i_n),\ i_j = 0, 1, \ldots, m_j - 1;\ j = 1, \ldots, n;\ i = 1, \ldots, v.$$

Assume further that the treatments are arranged lexicographically, as explained in Chapter 5.

A generalized cyclic design is generated from an initial block containing k distinct treatments by cyclically developing the contents of the initial block. Here each of the k-treatments in the initial block is a treatment combination represented by an n-tuple. If $\{(a_1, \ldots, a_n)\}$ denotes the set of k-treatments in the initial block, then the set of treatments of the i-th block is obtained by adding the i-th treatment combination $(i_1, \ldots, i_n)$ to the elements of the first set and the set of treatments of the i-th block will be $\{(c_1, \ldots, c_n)\}$, where a typical element of the i-th set is

$$(a_1, \ldots, a_n) + (i_1, \ldots, i_n) = (c_1, \ldots, c_n)$$

where

$$a_j + i_j \equiv c_j \bmod (m_j),\ j = 1, \ldots, n.$$

To illustrate the idea, consider an example of $v = 3 \times 2$ treatments arranged in blocks of size 2 each. Let the six treatment combinations be written as (0 0), (0 1), (1 0), (1 1), (2 0), (2 1). Let the initial block be (0 0), (1 1). The six blocks obtained by cyclic development of this initial block are

Blocks	Column–I	Blocks	Column–II
1.	(0 0), (1 1)	2.	(0 1), (1 0)
3.	(1 0), (2 1)	4.	(1 1), (2 0)
5.	(2 0), (0 1)	6.	(2 1), (0 0).

The GG/n designs generated through the procedure just described will clearly consist of v blocks although some blocks will have a fraction, say $1/p$, of the v blocks replicated p times. These sets of v/p distinct blocks are called fractional sets.

Consider again the example described above. This particular design is resolvable in the sense that each treatment combination occurs once in each column. It is clear that if $k = m_j$ for some $j = 1,..., n$, a resolvable design of v blocks in k groups of (v/k) blocks each can be constructed. Such a design will have each of the integers $0, 1,..., m_j - 1$ occuring once in the j-th position of the treatment combinations in the initial block. A fraction of three blocks can be generated from the initial block {(0 0), (1 1)} and a fractional set of two blocks can be generated from the initial block {(0 0), (1 0), (2 1)}. The generated design is

Blocks	Column–I	Blocks	Column–II
1.	(0 0), (1 0), (2 1)	2.	(0 1), (1 1), (2 0)
3.	(1 0), (2 0), (0 1)	4.	(1 1), (2 1), (0 0)
5.	(2 0), (0 0), (1 1)	6.	(2 1), (0 1), (1 0).

In the first case the two columns are same while in the second case the three rows are the same. Consider now the design GC/2 with $v = 4 \times 2$ treatments arranged in blocks of size three obtained by cyclically developing the initial block {(0 0), (1 0), (3 1)}. The design is

Blocks				Blocks			
1.	(0 0)	(1 0)	(3 1)	2.	(0 1)	(1 1)	(3 0)
3.	(1 0)	(2 0)	(0 1)	4.	(1 1)	(2 1)	(0 0)
5.	(2 0)	(3 0)	(1 1)	6.	(2 1)	(3 1)	(1 0)
7.	(3 0)	(0 0)	(2 1)	8.	(3 1)	(0 1)	(2 0)

The blocks along the columns are obtained by cyclically developing mod (4) the first position of the treatment combinations in the initial block and the blocks along the rows are obtained by developing cyclically mod (2) the second position of the treatment combinations in the blocks in the first column.

Generalized cyclic designs are constructed by taking the full or fractional set individually or in combinations and are of size $(v, k, r, m_1, \ldots, m_n)$. Consider the dicyclic solution of the BIB design $v = b = 16$, $r = k = 6$, $\lambda = 2$ reported by Fisher and Yates (1963, p. 90). This design is generated from the initial block {(0 0), (0 1), (0 2), (1 0), (2 3), (3 0)}. Again, the BIB design $v = 25$, $b = 50$, $r = 8$, $k = 4$, $\lambda = 1$ generated through two initial blocks {(0 0), (0 1), (1 0), (4 4)} and {(0 0), (0 2), (2 0), (3 3)} has also been reported by Fisher and Yates (1963). These two designs are GC/2 designs (16, 6, 6, 4, 4) and (25, 4, 8, 5, 5). Various group divisible, latin square type and cubic designs reported by John (1973) are generalized cyclic designs.

If v and m_j have some common divisor p_j $(j = 1, \ldots, n)$, then a fractional set of size $(v, k, k/p, m_1, \ldots, m_n)$ exists for each p, where $p = \prod_{j=1}^{n} p_j$.

6.7. FACTORIAL STRUCTURE IN GC/n DESIGNS

GC/n designs are a very useful class of designs for factorial experimentation. For these designs also, **NN'**, **A** and **Ω** are all circulant. Let $\mathbf{R}_n^i$ be an $n \times n$ circulant matrix whose first row has unity in the i-th column and zero everywhere else. The **Ω** matrix of GC/n designs can be written as

$$\mathbf{\Omega} = \sum_{i_1} \ldots \sum_{i_n} W_{i_1 \ldots i_n} \left(\prod_j \otimes \mathbf{R}_{m_j}^{i_j}\right) \qquad (6.1)$$

where

$$\prod_j \otimes \mathbf{R}_{m_j}^{i_j} = \mathbf{R}_{m_1}^{i_1} \otimes \ldots \otimes \mathbf{R}_{m_n}^{i_n}.$$

For GC/n designs, if the v treatments are identified with the factorial treatment combinations, then (6.1) can be written as

$$\mathbf{\Omega} = \sum_{i_1=1}^{m_1} \ldots \sum_{i_\alpha=1}^{m_\alpha} \left\{\left(\prod_{j=1}^{\alpha} \otimes \mathbf{R}_{m_j}^{i_j}\right) \otimes \mathbf{\Omega}_{i_1 \ldots i_\alpha}\right\}, \qquad (6.2)$$

where $\alpha = n-1$, and

$$\Omega_{i_1 \ldots i_\alpha} = \sum_{i_n=1}^{m_n} (\mathbf{R}_{m_n}^{i_n} \otimes W_{i_1 \ldots i_n}).$$

Clearly $\Omega_{i_1 \ldots i_\alpha}$ is a circulant matrix of order $m_n \times m_n$ with row and column sums equal. Comparing (6.2) with (5.23a), it follows that all GC/n designs have factorial structure and the analysis developed in the previous Chapter becomes directly applicable. This interesting property of GC/n designs, namely the orthogonal partitioning of the treatment sum of squares, has led to the studies of identifying confounded patterns in factorial experiments.

John (1973) has also worked out the average efficiency of a generalized interaction for a GC/n under factorial set up. While examining various designs John observed that the maximum main-effect efficiency is obtained for designs having $f_i(j)$ [equal or almost equal, where $f_i(j)$ denotes the frequency with which digit j occurs in the i-th position of a treatment in the initial block and satisfy

$$\sum_{j=0}^{m_i-1} f_i(j) = k, \quad i = 1, \ldots, n. \tag{6 3}$$

The above observation led John to conjecture that the maximum main-effect efficiency, for say the first factor, is obtained from the set of designs having s values of $f_1(i)$ equal to $[k/m_1]$ and the remaining $m_1 - s$ values of $f_1(i)$ equal to $[k/m_1] + 1$, where s is chosen so that (6.3) is satisfied and $[k/m_1]$ is the integer part of k/m_1. A similar result will hold for the main-effect efficiencies of the remaining $(n-1)$ factors.

Following John, Wolock and David (1972), the efficiency of the main-effect for the factor i has been obtained by John (1973) as

$$E_j = (m_j - 1)/rc, \tag{6.4}$$

where

$$c = \sum_{h=2}^{m_j} (1/\theta_h) \tag{6.5}$$

$$\theta_h = r - (1/k) \sum_{p=1}^{m_j} n_p^* \cos \{(h-1)(p-1)\phi\}, \tag{6.6}$$

$$h = 2, \ldots, m_j,$$

$$\phi = 2\pi/m_j,$$

and $n_p^* = n_{m_j}^* - p + 2$ is given by

$$n_p^* = \sum_{j=0}^{m_i-1} f_i(j) f_i(j+p-1), \tag{6.7}$$

with $j + p - i$ being reduced mod (m_i), wherever necessary. n_p^* is the p-th element of the first row of the circulant matrix $\mathbf{N}^*\mathbf{N}^{*\prime} = \{n_1^*, \ldots, n_{mj}^*\}$, where $\mathbf{N}^* = \mathbf{P}^x\mathbf{N}$ and $\mathbf{P}^x = \mathbf{I}_{mj} \otimes \mathbf{1}'_{v^*}$, $v^* = \prod_{j=2}^{n} m_j$.

The above results are useful in identifying partially confounded designs that have maximum or near maximum efficiency on all or some of the desired main-effects.

Example 6.1: We consider the example of 4^2 factorial experiments reported by John (1973) and examine the following initial blocks of three different designs:

$$\{(0\ 0), (0\ 1), (1\ 0), (1\ 1), (2\ 2), (3\ 3)\},$$

$$\{(0\ 0), (0\ 1), (1\ 0), (1\ 2), (2\ 1), (3\ 3)\},$$

$$\{(0\ 0), (0\ 1), (1\ 0), (1\ 2), (2\ 3), (3\ 1)\}.$$

The main-effect efficiencies of all the three designs as calculated from (6.4) are 0.9623. Any choice between these designs can therefore be made only on the basis of the interaction efficiencies. The efficiencies of interaction are 0.8028, 0.8247 and 0.8308 respectively, rendering the third design as better than the first two.

John and Dean (1975) and Dean and John (1975) have constructed single replicate generalized cyclic designs from a set of generating treatments. These are then used to provide a fairly general method of confounding in factorial experiments, symmetrical or otherwise. Their method of constructing single replicate GC/n designs is as follows:

As stated earlier, let the $v = \prod_{j=1}^{n} m_j$ treatments be denoted by the n-tuple $i = (i_1, \ldots, i_n)$, $i_j = 0, 1, \ldots, m_j - 1$; $j = 1, \ldots, n$, and $i = 1, \ldots, v$. Addition of treatment combinations is as defined in Section 6.6. Let the v treatment combinations form an abelian group, G, closed under addition. The GC/n design is obtained from an initial block whose treatment combinations form a subgroup S, of size k, from G. The remaining blocks are given by the cosets of S.

Let ϕ be the lowest common multiple of $m_1, \ldots, m_n$ and let

$$t = HCF(\phi, \phi i_1/m_1, \phi i_2/m_2, \ldots, \phi i_n/m_n), \tag{6.8}$$

where HCF denotes the highest common factor. If the treatment combinations obtained by multiplying each i_j in i by u, mod (m_j), is denoted by u_i, a cyclic abelian group S of size μ/t is generated by the combinations 0, i, $2i, \ldots, (\mu/t - 1)i$ because $\mu i/t \equiv 0$. The treatment combination i is the generator and the combinations in S constitute the initial block of a GC/n design with block size $k = \phi/t$.

When there are p generators $b_1, \ldots, b_p$ with corresponding abelian groups $S_1, \ldots, S_p$ such that the interaction of S_i and the direct sum of other groups is empty for all i, then the group S formed by taking the direct sum of $S_1, \ldots, S_p$ with general element

$$u_1 b_1 + \ldots + u_p b_p \; (u_i = 0, 1, \ldots, q_i - 1) \qquad (6.9)$$
$$i = 1, \ldots, p$$

has q distinct elements. Here t_i's are defined as in (6.8) for the generators b_i, $q_i = \phi/t_i$ and $q = \prod_{i=1}^{p} q_i$.

Example 6.2: Let $m_1=3$, $m_2=4$, and $m_3=6$, $b_1=(1\ 1\ 1)$ and $b_2=(0\ 2\ 3)$. We get $\phi = 12$, $t_1 = HCF(12, 4, 3, 2) = 1$ and $t_2 = HCF(12, 0, 6, 6) = 6$. The two generators therefore give a single replicate GC/n design for $v = 72$ and $k = 24$.

John and Dean have obtained simple expressions for calculating the number of degrees of freedom of the generalized interaction a^x confounded with the blocks and have used these to determine the confounding pattern of these designs. Dean and Lewis (1980) have further strengthened their results by presenting a unified theory for generalized cyclic designs. They have used the properties of abelian groups and have obtained necessary and sufficient conditions for the generation of fractional sets of blocks and also for the generation of resolvable designs.

7. Other Useful Incomplete Block Designs

In this Chapter we study several other incomplete block designs which are not only useful but also have close connections with the designs studied in the earlier Chapters.

7.1 DUAL DESIGNS

It has been mentioned in Chapter 2 that from a given incomplete block design many useful designs can be obtained. One such design is the dual design obtained from a given design by interchanging in it the roles of treatments and blocks.

Definition 7.1: If **N** is the incidence matrix of a block design D, then the design D^* with the incidence matrix **N′** is called the dual design.

The advantages of dual designs are many. Dualization of known designs may sometimes lead to new designs. Several of the two associate PBIB designs listed in Clatworthy (1973) are obtained as dual of known designs. In many cases dualization may also help in obtaining a simplified analysis by identifying the dual as a PEB design. Dual designs have also found their applications as mating designs in breeding experiments. For example, in the line $\times$ tester trials, the lines are identified as treatments and testers are blocks, and interest lies in making both the line and the tester comparisons with desired efficiency. The knowledge of the pattern of both the original design and its dual design is essential for such mating designs.

7.1.1 Some earlier developments

Bose and Nair (1939) gave examples of PBIB designs obtained by dualizing some BIB designs. Youden (1951) investigated these further. Shrikhande (1952) and Shrikhande and Bhagwandas (1965) studied the dual of asymmetrical BIB and two associate class PBIB designs. They showed that the dual of all BIB designs with $\lambda = 1$ and some BIB designs with $\lambda = 2$ are PBIB designs with two associate classes. Utilizing the result that the dual of a partial geometry (r, k, t) is again a partial geometry (k, r, t), they showed that the duals of PBIB designs with two associate classes constructed from a partial geometry are again PBIB designs with two associate classes. They have also shown that the dual of a symmetrical two associate class PBIB design with $\lambda_1 = \lambda_2 \pm 1$ is again a PBIB design. Further, the dual of a three associate class PBIB design, with λ_i ($i = 1, 2, 3$) either 0 or 1, $v > b$ and concurrence matrix non-singular, is a two associate PBIB design. Ramakrishnan (1956) obtained a new series of five associate class PBIB designs with two replications by dualizing a series of group divisible designs with $k = 2$ and has shown that the analysis of the dual designs is extremely simple. Shrikhande and Bhagwandas (1965) showed that the dual of an affine α-resolvable PBIB design is a semi-regular group divisible design. The work reported here is, however, restricted to equi-replicate, proper and binary designs.

Given two designs D and D^* it may be of interest to know whether the design D^* is a dual of D or not. If D^* is a dual of D then it is necessary that the number of treatments in D^* be equal to the number of blocks in D, the number of blocks in D^* be equal to the number of treatments in D and the number of replications of treatments in D^* be equal to the block sizes in D. Further, if $\mathbf{N}$ is the incidence matrix of D and $\mathbf{N}^*$ is the incidence matrix of D^* then the non-zero distinct eigen values $\mu_0, \mu_1, \ldots, \mu_m$ alongwith their multiplicities should be the same for $\mathbf{NN}'$ and $\mathbf{N}^*\mathbf{N}^{*\prime}$. Furthermore, in order that D^* be a dual of D, it is necessary that the gramian P_i, of the independent, rational, eigen vectors corresponding to the eigen values μ_i, of $\mathbf{N}^*\mathbf{N}^{*\prime}$ be rationally congruent to Q_i^*, the gramian of the independent rational eigen vectors corresponding to the eigen values μ_i of $\mathbf{N}'\mathbf{N}$. Using this result Raghavarao (1966) obtained necessary conditions for the dual of a symmetrical PBIB design with the association scheme say A, to be a PBIB design with the same set of parameters and the same association scheme A. Hoffman (1963) obtained conditions under which the dual of a regular, symmetric PBIB design with m associate classes is again a PBIB design with the same set of parameters, whatever be the association scheme.

7.1.2 Some recent developments

We now present some further results on dual designs. We shall first give the following Lemma [See, Roy and Laha (1956)] which shall be referred to in all the theorems that follow:

Lemma 7.1: If $\mathbf{A}$ and $\mathbf{B}$ are two matrices such that $\mathbf{AB}$ and $\mathbf{BA}$ are defined then the non-zero eigen values of $\mathbf{AB}$ and $\mathbf{BA}$ are identical and have the same multiplicities. Further, if corresponding to an eigen value μ, $\boldsymbol{\xi}$ is an eigen vector of $\mathbf{AB}$ then $\boldsymbol{\eta} = \mathbf{B}\boldsymbol{\xi}$ will be the eigen vector of $\mathbf{BA}$ corresponding to the same eigen value μ. The following theorem is due to Ceranka and Mejza (1978) and Kageyama (1980). It is assumed that the original design is connected.

Theorem 7.1: Let $D(v, b, \mathbf{r}, \mathbf{k})$ be an efficiency balanced (EB) design. The dual design D^* is SPEB if $b > v$ and EB if $b = v$.

Proof: By definition, M-matrix of D is

$$\mathbf{M} = (\mathbf{R}^{-1}\mathbf{N})\,(\mathbf{K}^{-1}\mathbf{N}')$$

and that of D^* is

$$\mathbf{M}^* = (\mathbf{K}^{-1}\mathbf{N}')\,(\mathbf{R}^{-1}\mathbf{N}).$$

Taking $\mathbf{R}^{-1}\mathbf{N} = \mathbf{A}$ and $\mathbf{K}^{-1}\mathbf{N}' = \mathbf{B}$, we can express $\mathbf{M}$ as $\mathbf{AB}$ and $\mathbf{M}^*$ as $\mathbf{BA}$. As D is efficiency balanced, $\mathbf{M}$ has only one non-zero eigen value μ with multiplicity $v - 1$ (other than the eigen value 1 corresponding to the eigen vector $\mathbf{1}$). Utilizing the result of Lemma 7.1, $\mathbf{M}^*$ also has the non-zero eigen values as μ with multiplicity $v - 1$. But as $\mathbf{M}^*$ has the order $b \times b$ it must have eigenvalue 0 (other than corresponding to the eigen vector $\mathbf{1}$) with multiplicity $b - v$. The design D^* is therefore SPEB if $b > v$ and EB if $b = v$. This proves the theorem.

Corollary 7.1: For EB designs, $b \geqslant v$. Since $b - v$ is the multiplicity of the eigen value 0 it must be non-negative. Hence $b - v \geqslant 0$ which implies that $b \geqslant v$. As a BIB design is trivially EB, we have an alternative proof of the Fisher's inequality, viz, $b \geqslant v$, for a BIB design.

Corollary 7.2: The dual of a symmetrical BIB design is a symmetrical BIB design and the dual of a non-symmetrical BIB design is SPEB.

The second part of the result of Corollary 7.2 is important as it reveals the structure of the dual of all asymmetrical BIB designs. The earlier work of Shrikhande and Bhagwandas identifying the dual of a BIB design as a two associate class PBIB design was available only when the original BIB design has $\lambda = 1$ or 2. Thus utilizing the result of Corollary 7.2, a large number of useful dual designs can be derived which have a simplified analysis.

We now prove a theorem due to Nigam and Puri (1982).

Theorem 7.2: The dual of a SPEB is SPEB if $\rho < b - 1$ and is EB if $\rho = b - 1$, where ρ is the multiplicity of the non-zero eigen values of $\mathbf{M}$ matrix.

Proof: Let D be a SPEB design with parameters v, b, $\mathbf{r}$, $\mathbf{k}$, μ, $\mathbf{L}$, where μ is the non-zero eigen value of $\mathbf{M}$ with multiplicity $\rho < v - 1$ and $\mathbf{L}$ be the corresponding idempotent matrix of rank ρ. The non-zero eigen value of $\mathbf{M}$ of D and of $\mathbf{M}^*$ of D^*, the dual of D, is same μ with the same multiplicity ρ. As $\mathbf{M}^*$ is of order $b \times b$, D^* is SPEB if $\rho < b - 1$ and is EB if $\rho = b - 1$.

If $\mathbf{s}_i$ is the set of eigen vectors, $i = 1, \ldots, \rho$, of $\mathbf{M}$ corresponding to the non-zero eigen value μ with multiplicity ρ, satisfying $\mathbf{M}\mathbf{s}_i = \mu\,\mathbf{s}_i$, then using part II of Lemma 7.1 and noting that $\mathbf{M}=\mathbf{R}^{-1}\mathbf{N}\mathbf{K}^{-1}\mathbf{N}'$, it is easily seen that the right eigen vectors of $\mathbf{M}^*$ corresponding to the non-zero eigen values μ will be

$\mathbf{s}_i^* = (\mathbf{K}^{-1}\mathbf{N}')\,\mathbf{s}_i$, $i = 1,\ldots,\rho$, and the idempotent matrix $\mathbf{L}^*$ will be

$$\mathbf{L}^* = \sum_{i=1}^{\rho} \mathbf{s}_i^*\,\mathbf{s}_i^{*\prime}\mathbf{K}/\mathbf{s}^{*\prime}\,\mathbf{K}\,\mathbf{s}_i^*$$

$$= (\mathbf{K}^{-1}\mathbf{N}')\,(\sum_{i=1}^{\rho} \mathbf{s}_i\mathbf{s}_i'\mathbf{R})\,(\mathbf{R}^{-1}\mathbf{N})/\mathbf{s}_i'\mathbf{R}\mathbf{s}_i\,\mu$$

$$= (1/\mu)\,(\mathbf{K}^{-1}\mathbf{N}')\,\mathbf{L}\,(\mathbf{R}^{-1}\,\mathbf{N}).$$

The above holds because $\mathbf{s}_i^{*\prime}\,\mathbf{K}\mathbf{s}_i^* = \mathbf{s}_i'\,\mathbf{N}\mathbf{K}^{-1}\mathbf{N}'\mathbf{s}_i = \mathbf{s}_i'\mathbf{R}\mathbf{M}\mathbf{s}_i = \mathbf{s}_i'\mathbf{R}\mu\,\mathbf{s}_i = \mu\mathbf{s}_i'\mathbf{R}\mathbf{s}_i$.

Corollary 7.3: The dual of every two associate class PBIB design having only one non-zero eigen value of $\mathbf{NN}'$ is SPEB.

Clearly if D has parameters $(v, b, r, k, \mu, \mathbf{L})$ then D^* will have the parameters $v^* = b$, $b^* = v$, $r^* = k$, $k^* = r$, $\mu^* = \mu$, $\mathbf{L}^* = \mathbf{N}'\mathbf{L}\mathbf{N}/\theta$ where $\mu = \theta/rk$, θ being the non-zero eigen value of $\mathbf{NN}'$ with multiplicity $\rho < v - 1$. The result contained in Corollary 7.3 is useful as it reveals that the duals of affine-resolvable PBIB, singular and semi-regular GD, simple lattice, PBIB obtained through partial geometry (r, k, t) are equi-replicate and proper SPEB designs. The analysis of all such dual designs can be simplified by analysing these as SPEB designs. For such D^*, we have

$$\mathbf{M}_0^* = \mu\,[\mathbf{N}'\mathbf{N}/\mu rk - \mathbf{1}\mathbf{1}'/b\mu] = \mu\mathbf{L}^* \text{ say}$$

and

$$\boldsymbol{\Omega}^* = [\mathbf{I} + \mathbf{N}'\mathbf{N}/r\,(1-\mu)\,k - \mathbf{1}\mathbf{1}'/b\,(1-\mu)]/k$$

and

$$\mathbf{Q}^{*\prime}\boldsymbol{\Omega}^*\mathbf{Q}^* = \mathbf{Q}^{*\prime}\mathbf{Q}^*/k + (\mathbf{N}\mathbf{Q}^*)'\,(\mathbf{N}\mathbf{Q}^*)/rk^2\,(1-\mu)$$

$$= \Sigma\, Q_i^{*\,2}/k + \Sigma\, u_j^{*\,2}/rk^2\,(1-\mu), \text{ where}$$

Q_i^* and u_j^*, $i, j = 1,\ldots, b$ are the elements of $\mathbf{Q}^*$ and $\mathbf{U}^* = \mathbf{N}\mathbf{Q}^*$ respectively, and Q_i^*'s are the adjusted treatment totals of the dual design D^*.

We now prove a more general theorem giving dual of PEB designs.

Theorem 7.3: The dual of a PEB design with m efficiency classes is a PEB design with at most $(m + 1)$ efficiency classes.

Proof: Since D is PEB (m), the eigen values of $\mathbf{M}$ are μ_i, $i = 1,\ldots, m$ with multiplicities ρ_i such that $\Sigma\,\rho_i = v - 1$. Let all μ_i, $i = 1,\ldots, m$, be non-zero. When $b > v$, the eigen values of $\mathbf{M}^*$ of the dual design will also be μ_i, $i = 1,\ldots, m$ and $\mu_{m+1} = 0$ with respective multiplicities ρ_i and $\rho_{m+1} = b - v$. If $b = v$, the eigen values of $\mathbf{M}^*$ will be μ_i, $i = 1,\ldots, m$ with multiplicities

ρ_i. As **M** and **M*** have the same non-zero eigen values with same multiplicities, it is obvious that if **M** has all non-zero eigen values, then b cannot be less then v. The dual design is therefore PEB $(m+1)$ or PEB (m) according as $b > v$ or $b = v$.

Consider now the case when one of the eigen values of **M**, say μ_m, with multiplicity ρ_m, is zero. If $b \geqslant v$, **M*** will also have $m-1$ non-zero eigen values μ_i with multiplicities ρ_i, $i = 1,\ldots, m-1$ and a zero eigen value with multiplicity $\rho_m^* = \rho_m + b - v$. When $b = v$, $\rho_m^* = \rho_m$. Thus for the case $b \geqslant v$, the dual design is again PEB(m). When $b < v$, **M*** will have $m-1$ non-zero eigen values with multiplicities ρ_i, $i = 1,\ldots, m-1$ such that $\sum_{i=1}^{m-1} \rho_i = v-1-\rho_m$, where ρ_m is the multiplicity of zero eigen value of **M** of the design D. But as **M*** has $b-1$ eigen values, the $b-1-(v-1-\rho_m) = b-v+\rho_m$ eigen values are zero. Thus for the case $b < v$, the dual design is again PEB(m).

Following on the lines of Corollary 7.3, it can easily be shown that the matrices $\mathbf{L}_i^*$ of D^* are

$$\mathbf{L}_i^* = \mathbf{K}^{-1}\mathbf{N}'\mathbf{L}_i\mathbf{R}^{-1}\,\mathbf{N}/\mu_i,\ \mu_i > 0,$$

where $\mathbf{L}_i$ are the idempotent matrices corresponding to the design D.

The result contained in Theorem 7.3 is again very useful as it follows from it that the dual of a PBIB design with m associate classes is PEB with at most $(m+1)$ efficiency classes. This result has wider applicability and is far more general than the earlier results known on the duals of PBIB designs. As an instance the five associate class PBIB designs of Ramakrishnan (1956) belong to this class as they are the dual of group divisible PBIB designs with two associate classes. His five associate class PBIB designs are as a matter of fact PEB(3) with one of the eigen values zero and therefore can be analysed easily.

The dual designs lead to another class of useful designs called Linked Block (LB) designs. These designs were introduced by Youden (1951) who obtained these as duals of BIB designs. The LB design so constructed by Youden happened to be all PBIB designs, but this is not necessarily true. We shall see later that LB designs are necessarily SPEB and as such their analysis gets simplified.

7.2 LINKED BLOCK DESIGNS

Definition 7.2: A binary incomplete block design is said to be of a Linked Block (LB) type if any two blocks in the design have a constant number of treatments common.

It follows from the definition that if **N** is the incidence matrix of a LB design, then we must have

$$\mathbf{N}'\mathbf{N} = \mathbf{d}^{\delta} + \phi\mathbf{1}\mathbf{1}' \tag{7.1}$$

where $\mathbf{d}^\delta$ is a diagonal matrix with diagonal elements $\mathbf{d}$ and ϕ is a scalar. For a general block design $D(v, b, \mathbf{r}, \mathbf{k})$ it implies from the condition (7.1) that the dual design must be pairwise balanced. Clearly, for such designs $b \geqslant v$.

Roy and Laha (1956) classified the Linked Block Designs into the following categories:

i) Symmetrical BIB designs
ii) PBIB designs
iii) Irregular designs not belonging to any of the known types.

The classification of Roy and Laha and other earlier work are restricted to proper and equi-replicate designs only. We now prove the following theorem due to Puri and Nigam (1982) for a general block design $D(v, b, \mathbf{r}, \mathbf{k})$.

Theorem 7.4: Consider an incomplete block design $D(v, b, \mathbf{r}, \mathbf{k})$.

i) If D is pairwise balanced, then its dual is always a LB type design.
ii) If D is EB design, then its dual is of LB type if D is equi-replicate and proper.
iii) If D is variance balanced design, then its dual is linked block if D is proper.

The theorem can easily be proved by making an appeal to the characterizations (3.45), (3.33) and Remark 3.8 given in Chapter 3.

Theorem 7.5: The necessary and sufficient condition for an equi-replicate and proper design $D(v, b, r, k)$ to be of LB type is that $k-\phi$ is an eigen value of the treatment structure matrix $\mathbf{NN}'$ with multiplicity $(b-1)$, where $\phi = k(r-1)/(b-1)$.

Proof: Before we give the proof of the theorem we present a Lemma which shall be useful in the proof.

Lemma 7.2: The necessary and sufficient condition for a symmetric matrix $\mathbf{A}$ of order n to have all its diagonal elements equal and all its off diagonal elements equal is that it has only two eigen values, one with multiplicity $(n-1)$ and the other corresponding to the eigen vector $\mathbf{1}$.

The proof of the theorem is as follows. Using Lemma 7.2 the necessary part of the theorem is obvious. To prove the sufficiency, let $\mathbf{N}$ be the incidence matrix of the given design $D(v, b, r, k)$. In order that $\mathbf{N}$ is the incidence matrix of a LB design, the block-structure matrix $\mathbf{N'N}$ should have all off-diagonal elements equal. Since it is given that $k-\phi$ is an eigen value of $\mathbf{NN}'$ with multiplicity $(b-1)$, then by Lemma 7.1, $\mathbf{N'N}$ will also have

$k-\phi$ as its non-zero eigen value with multiplicity $(b-1)$. Since $\mathbf{NN'1}=rk\mathbf{1}$, rk is an eigen value of $\mathbf{NN'}$ with corresponding eigen vector $\mathbf{1}$. Also, rk is an eigen value of $\mathbf{N'N}$ corresponding to the eigen vector $\mathbf{1}$. Thus $\mathbf{N'N}$ satisfies all the conditions of Lemma 7 2, and hence its off diagonal elements are equal. The diagonal elements of $\mathbf{N'N}$ are k and off-diagonal elements are ϕ. This completes the proof.

Corollary 7.4: If the treatment structure matrix $\mathbf{NN'}$ of any design $D(v, b, r, k)$ has only two non-zero eigen values, rk and $k(b-r)/(b-1)$, and rk is not a repeated eigen value, then the design must be linked block.

Proof: Let ρ be the multiplicity of the eigen value $k(b-r)/(b-1)$. We have

$$\text{Trace of } \mathbf{NN'} = \text{Sum of eigen values of } \mathbf{NN'}$$
$$= rk + \rho k(b-r)/(b-1)$$

But trace of $\mathbf{NN'} = rv$. Thus

$$rv = rk + \rho k(b-r)/(b-1).$$

Using $\quad vr = bk$, we get $\rho = b-1$.

Corollary 7.5: If $D(v, b, r, k)$ is variance balanced, then a necessary and sufficient condition for D to be of the LB type is that $b = v$.

Proof: Since the design is variance balanced, $\mathbf{NN'}$ has all its non-zero eigen values equal with multiplicity $(v-1)$, while if it is of LB type the multiplicity must be $(b-1)$. Hence $v = b$.

It follows from Corollary 7.4 that an equi-replicate and proper SPEB design is always of LB type if θ_1, the non-zero eigen value of $\mathbf{NN'}$ with multiplicity $(b-1)$, equals $k(b-r)/(b-1)$. Hence all LB type designs are necessarily SPEB and can be analysed as SPEB designs. This identification leads to simplified analysis of LB designs. We can easily obtain the necessary and sufficient condition for a SPEB type PBIB design to be of LB type by equating the non-zero eigen values of $\mathbf{NN'}$ of the PBIB design to $k(b-r)/(b-1)$. Thus, the semi-regular GD will be of LB type if $k(b-r)/(b-1) = r-\lambda_1$, which on solving gives $b = v-m+1$. This condition was obtained earlier by Roy and Laha (1957) through a different and a longer procedure. The approach given here is due to Nigam and Puri (1982). Following on similar lines it is easy to prove that a singular GD is of LB type iff $b = m$, an L_i PBIB is of LB type iff $r = (i-s)(\lambda_1-\lambda_2)$ and $b = s(s-i)+i$ or $r = i(\lambda_1-\lambda_2)+\lambda_2$ and $b = i(s-1)+1$, a triangular design is of LB type iff $r = 2\lambda_1-\lambda_2$ and $b = n$ or $r = (n-3)\lambda_2-(n-4)\lambda_1$ and $b = i(s-1)+1$.

7.3 SUPPLEMENTED BLOCK DESIGNS

In plant breeding experiments with new strains or in bio-chemical research with new pesticides, herbicides, soil fumigants, drugs, etc., the major problem is the evaluation of new strains or chemicals. In such comparative experiments generally limited material is available for the new strains or chemicals (treatments) as compared to standard (or check) treatments. The material for new treatments is so meagre that it is just sufficient for making only one or two replications. Reinforced (Das, 1958), Augmented (Federer, 1956, 1961), Supplemented balanced (Pearce, 1960), Nearly Balanced (Nigam, 1976), Orthogonally Supplemented balanced (Calinski, 1971), Supplemented block (Puri, Nigam and Narain, 1977; Singh and Dey, 1979) designs have been developed for such experiments. The basic approach in these designs is to augment (supplement) any standard design in check (or new) treatments with a set of new (or check) treatments. The augmentation is done by including additional treatments in each block of the standard design. Since all the designs developed meet more or less the same objectives and as there is considerable overlap between supplemented block and other class of designs referred above, we discuss here only supplemented block designs. For details, reference may be made to Nigam, Gupta and Narain (1981).

Definition 7.3: Consider a connected block design D $(v, b, \mathbf{r}_1, \mathbf{k}_1)$ with incidence matrix $\mathbf{N}_1$ and $\mathbf{r}_1{}'\,\mathbf{1} = \mathbf{k}_1{}'\mathbf{1} = n_1$. Supplement each of the b blocks of D with new treatments so that the supplementary treatments form a design $D_1(v_1, b, \mathbf{r}_2, \mathbf{k}_2)$ with incidence matrix $\mathbf{N}_2$ and $\mathbf{r}_2{}'\mathbf{1} = \mathbf{k}_2{}'\mathbf{1} = n_2$. The resulting design D^* with incidence matrix

$$\mathbf{N}^* = \begin{bmatrix} \mathbf{N}_1 \\ \mathbf{N}_2 \end{bmatrix}$$

is called a supplemented block design. The parameters of the supplemented block design are

$$v^* = v + v_1,\ b^* = b,\ \mathbf{r}^* = [\mathbf{r}_1{}' : \mathbf{r}_2{}']',\ \mathbf{k}^* = \mathbf{k}_1 + \mathbf{k}_2. \tag{7.2}$$

We now describe the analysis of supplemented block designs.

7.3.1 Analysis of supplemented block designs

This Section is devoted to the general analysis of supplemented block designs. It will be seen that this analysis does not pose any new problems and the analysis follows on the lines of the general analysis of block designs. The particular cases of interest have been dealt with separately.

Consider a supplemented block design D^* with parameters as given in (7.2) and incidence matrix

$$\mathbf{N}^* = \begin{bmatrix} \mathbf{N}_1 \\ \mathbf{N}_2 \end{bmatrix}.$$

The matrix $\mathbf{M}$ of the design D^* is

$$\mathbf{M}^* = \begin{bmatrix} \mathbf{R}_1^{-1}\mathbf{N}_1\mathbf{K}^{-1}\mathbf{N}_1' & \mathbf{R}_1^{-1}\mathbf{N}_1\mathbf{K}^{-1}\mathbf{N}_2' \\ \mathbf{R}_2^{-1}\mathbf{N}_2\mathbf{K}^{-1}\mathbf{N}_1' & \mathbf{R}_2^{-1}\mathbf{N}_2\mathbf{K}^{-1}\mathbf{N}_2' \end{bmatrix},$$

where $\mathbf{R}_j$ is a diagonal matrix with diagonal elements $\mathbf{r}_j$, $j = 1, 2$, and $\mathbf{K}$ is a diagonal matrix with diagonal elements $\mathbf{k}^*$. The matrix $\mathbf{M}_0$ of the design D^* can be obtained as

$$\mathbf{M}_0^* = \begin{bmatrix} \mathbf{R}_1^{-1}\mathbf{N}_1\mathbf{K}^{-1}\mathbf{N}_1' - \mathbf{1}\mathbf{r}_1'/n & \mathbf{R}_1^{-1}\mathbf{N}_1\mathbf{K}^{-1}\mathbf{N}_2' - \mathbf{1}\mathbf{r}_2'/n \\ \mathbf{R}_2^{-1}\mathbf{N}_2\mathbf{K}^{-1}\mathbf{N}_1' - \mathbf{1}\mathbf{r}_1'/n & \mathbf{R}_2^{-1}\mathbf{N}_2\mathbf{K}^{-1}\mathbf{N}_2' - \mathbf{1}\mathbf{r}_2'/n \end{bmatrix} \quad (7.3)$$

where $n = n_1 + n_2$ is the total number of observations in D^*.

If the supplementary treatments added to the blocks of D are in number proportional to the original block sizes, i.e. $\mathbf{k}^* = n\,\mathbf{k}_1/n_1$ and $\mathbf{N}_2 = \mathbf{r}_2\,\mathbf{k}^{*\prime}/n$ as postulated for the design under consideration, because the proportionality factor may not be equal for all the supplementary treatments and may depend upon their number of replications, then $\mathbf{M}_0^*$ in (7.3) can be shown to be equal to

$$\mathbf{M}_0^* = \begin{bmatrix} n_1/n\mathbf{M}_0 & \mathbf{0} \\ \mathbf{0} & \mathbf{0} \end{bmatrix} = \sum_{i=1}^{m} \mu_i^* \mathbf{L}_i^*, \quad (7.4)$$

where

$$\mathbf{M}_0 = \mathbf{R}_1^{-1}\mathbf{N}_1\mathbf{K}_1^{-1}\mathbf{N}_1' - \mathbf{1}\mathbf{r}_1'/n_1$$

is the usual $\mathbf{M}_0$ matrix of the design D and $\mathbf{K}_1$ is a diagonal matrix with diagonal elements $\mathbf{k}_1$. Further $\mu_i^* = n_1\mu_i/n$ and

$$\mathbf{L} = \begin{bmatrix} \mathbf{L}_i & \mathbf{0} \\ \mathbf{0} & \mathbf{0} \end{bmatrix}, \; i = 1, \ldots, m.$$

Also $\mathbf{M}_0 = \sum_{i=1}^{m} \mu_i \mathbf{L}_i$, if D is a PEB design with m efficiency classes.

Clearly, the eigen values of $\mathbf{M}_0^*$ are $\mu_1^*, \ldots, \mu_m^*, \mu_{m+1}^*$, with multiplicities $\rho_1, \rho_2, \ldots, \rho_m, \rho_{m+1}$ (the eigen value zero corresponding to the eigen vector $\mathbf{1}$ is not mentioned) and the corresponding associated idempotent matrices are

$$\mathbf{L}_1^*, \ldots, \mathbf{L}_m^*, \mathbf{L}_{m+1}^* = \mathbf{I} - \mathbf{1}\mathbf{r}^{*\prime}/n - \sum_{i=1}^{m} \mathbf{L}_i^*,$$

respectively. It can be shown easily that the set of $(v - 1)$ treatment contrasts $\mathbf{s}^*_{ij}$ given as

$$\mathbf{s}^*_i = \begin{bmatrix} \mathbf{s}_{ij} \\ \mathbf{0} \end{bmatrix}, i = 1,\ldots, m; j = 1,\ldots, \rho_i,$$

are eigen vectors of $\mathbf{M}^*_0$ corresponding to the eigen values μ^*_i and these contrasts are estimated with efficiency factor $(1 - \mu^*_i)$. Here $\mathbf{s}_{ij}$ are the basic contrasts of the design D having the efficiency factor $(1 - \mu_i)$. It may be noted that $(1 - \mu^*_i) > (1 - \mu_i)$. Thus the addition of supplementary treatments is likely to result in an increase in efficiency factor associated with contrasts of basic set of treatments in D provided that the within block variance of the design D and D^* are of the same order. It can be easily shown that (v_1-1) treatment contrasts among supplementary treatments and a contrast between supplementary and the basic set of treatments is estimated with efficiency factor one. We thus have the following theorem due to Puri, Nigam and Narain (1977):

Theorem 7.6: The design D^* is PEB with atmost $(m + 1)$-efficiency classes and parameters

$$v^* = v + v_1,\ b^* = b,\ \mathbf{r}^* = [\mathbf{r}_1'\mathbf{r}_2']',\ \mathbf{k}^* = n\mathbf{k}_1/n_1,$$

$$\mu^*_i = n_1\mu_i/n,\ \rho^*_i = \rho_i,\ \mathbf{L}^*_i,\ i = 1,\ldots, m + 1.$$

The $\mathbf{\Omega}$ matrix of D^* will be

$$\mathbf{\Omega}^* = [\mathbf{I} + \sum_{i=1}^{m} \{\mu^*_i/(1 - \mu^*_i)\}\, \mathbf{L}^*_i\,]\, \mathbf{R}^{-1}.$$

Let $\mathbf{Q}^*$ denote the vector of adjusted treatment totals for D. Then the sum of squares attributed to treatments is given by

$$\mathbf{Q}^{*\prime}\mathbf{\Omega}^*\mathbf{Q}^* = \mathbf{Q}'\, [\mathbf{I} + \sum_{i=1}^{m} \mu^*_i\, \mathbf{L}_i^*/(1 - \mu^*_i)]\, \mathbf{R}_1^{-1}\, \mathbf{Q}$$
$$+ \mathbf{Q}'_{v_1}\, \mathbf{R}_2^{-1}\, \mathbf{Q}_{v_1}$$

where $\mathbf{Q}$ and $\mathbf{Q}_{v_1}$ are the vectors of adjusted treatment totals corresponding to v basic set of treatments and v_1 supplementary treatments.

Nigam and Puri (1982) considered an m-associate resolvable PBIB design D with parameters v, b, r, k, λ_i, $i = 1,\ldots, m$, as the standard design. The blocks are grouped in r sets such that each set is a complete replicate. Add the supplementary treatment $v + 1$ to each block of first set, the treatment $v + 2$ to each block of second set, and so on, and the treatment $v + r$ to each block of the r-th set. Thus while the first v treatments are replicated r times, the v supplementary treatments are replicated b/r times. The incidence matrix of the resulting design D^* is

$$\mathbf{N}^* = \begin{bmatrix} \mathbf{N}_1 \\ \mathbf{I}_r \otimes \mathbf{1}'_{b/r} \end{bmatrix}.$$

The matrix $\mathbf{M}_0$ of the design D^* is

$$\mathbf{M}_0^* = \begin{bmatrix} \sum_{i=1}^{m} \mu^* \mathbf{L}_i & \mathbf{0} \\ \mathbf{0} & \mathbf{0} \end{bmatrix} + 1/(k+1) \begin{bmatrix} \mathbf{0} & \mathbf{0} \\ \mathbf{0} & \mathbf{I} - \mathbf{J}/\mathbf{r} \end{bmatrix}.$$

The eigen values of $\mathbf{M}_0^*$ are $\mu_i^* = k\mu_i/(k+1)$, $i = 1,\ldots, m$, and $\mu_{m+1}^* = 1/(k+1)$.

Singh and Dey (1979) gave the analysis of supplemented designs when the basic design D is either a BIB or a linked block design and the number of supplementary treatments added is equal to the number of blocks in D. These cases were also considered earlier by Federer and Raghavarao (1975).

7.3.2. Check treatments in BIB design

Let v check treatments be randomly allotted to the experimental units within each of b blocks of the BIB design and b new treatments be randomly assigned so that every new treatment appears only in one of the b blocks. The incidence matrix of the resulting design D^* is

$$\mathbf{N}^* = \begin{bmatrix} \mathbf{N}_1 \\ \mathbf{I} \end{bmatrix} \tag{7.5}$$

We then have the following theorem:

Theorem 7.7: The design with the incidence matrix (7.5) is PEB with three efficiency classes and the parameters

$v^* = v + b$, $b^* = b$, $\mathbf{r}^* = (r\,\mathbf{1}'\ \mathbf{1}')'$, $\mathbf{k}^* = (k+1)\,\mathbf{1}$,

$\mu_1^* = (2r - \lambda)/(rk + r)$, $\rho_1^* = v$, $\mu_2^* = 1/(k+1)$, $\rho_2^* = b - v$,

$\mu_3^* = 0$, $\rho_3^* = v - 1$,

$$\mathbf{L}_1^* = \{1/v(r-\lambda)(2r-\lambda)\} \begin{bmatrix} (r-\lambda)^2\,(v\mathbf{I} - \mathbf{11}') & (r-\lambda)\,(v\mathbf{N}_1 - k\mathbf{11}') \\ r(r-\lambda)\,(v\mathbf{N}_1 - k\mathbf{11}') & r(v\mathbf{N}_1{}'\mathbf{N}_1 - k^2\mathbf{11}') \end{bmatrix}$$

$$\mathbf{L}_2^* = \{1/(r-\lambda)\} \begin{bmatrix} \mathbf{0} & \mathbf{0} \\ \mathbf{0} & (r-\lambda)\mathbf{I} + (\lambda k/r)\mathbf{11}' - \mathbf{N}_1{}'\mathbf{N}_1 \end{bmatrix}$$

$$\mathbf{L}_3^* = \mathbf{I} - \mathbf{L}_1^* - \mathbf{L}_2^* - \mathbf{1r}^{*\prime}/n,$$

where $n = b\,(k+1)$.

The matrix $\mathbf{\Omega}$ for the design D^* is

$$\mathbf{\Omega}^* = \begin{bmatrix} (k/\lambda v)\mathbf{I} - (r-\lambda)\mathbf{1}\mathbf{1}'/\lambda r v^2 & (1/\lambda v^2)(v\mathbf{N}_1 - k\,\mathbf{1}\mathbf{1}') \\ (1/\lambda r v^2)(v\mathbf{N}_1' - k\mathbf{1}\mathbf{1}') & (1+k)\mathbf{I}/k + \mathbf{N}_1'\mathbf{N}_1/kv - (\lambda v + rk)\mathbf{1}\mathbf{1}'/\lambda v^2 \end{bmatrix}$$

and the adjusted treatment sum of squares is

$$\mathbf{Q}^{*\prime}\mathbf{\Omega}^*\mathbf{Q}^* = (k+1)\,\mathbf{Q}_2'\mathbf{Q}_2/k + k\,\mathbf{Q}_1'\mathbf{Q}_1/\lambda v + \mathbf{P}_2'\mathbf{P}_2/\lambda v k + 2\mathbf{Q}_2'\,\mathbf{P}_2/\lambda v,$$

where $\mathbf{P}_2 = \mathbf{N}_1\mathbf{Q}_2$, and $\mathbf{Q}_1$ and $\mathbf{Q}_2$ denote respectively the vector of adjusted totals of basic treatments and the supplementary treatments.

Corollary 7.6: The supplemented design with the incidence matrix

$$\mathbf{N}^* = \begin{bmatrix} \mathbf{1}\mathbf{1}' \\ \mathbf{I} \end{bmatrix}$$

is a SPEB with two efficiency classes and parameters
$v^* = v + b$, $b^* = b$, $\mathbf{r}^* = (b\mathbf{1}'\ \mathbf{1}')'$, $k^* = (k+1)\mathbf{1}$, $\mu_1^* = 1/(v+1)$, $\rho_1^* = b$, $\mu_2^* = 0$, $\rho_2^* = v - 1$,

$$\mathbf{L}_1^* = \begin{bmatrix} \mathbf{0} & \mathbf{0} \\ \mathbf{0} & \mathbf{I} - \mathbf{1}\mathbf{1}'/b \end{bmatrix}, \quad \mathbf{L}_2^* = \mathbf{I} - \mathbf{L}_1^* - \mathbf{1}\mathbf{r}^{*\prime}/n,$$

where $n = b(v+1)$. The adjusted treatment sum of squares is

$$(1 + 1/v)\,\mathbf{Q}_2'\mathbf{Q}_1 + \mathbf{Q}_1'\mathbf{Q}_1/b - S_2^2/bv,$$

where $S_2 = \mathbf{Q}_2'\,\mathbf{1}$. This is one of the cases considered by Federer (1961).

7.3.3 Check treatments in linked block design

Consider a BIB design with parameters v, b, r, k, λ. Corresponding to this there exists a linked block design, obtained by dualization of the BIB design, in b treatments and v blocks, k replications and block size r. If $\mathbf{N}_1$ is the incidence matrix of the BIB design then $\mathbf{N}_1'$ is the incidence matrix of the corresponding linked block design. Let v new treatments be introduced at the rate of one in each of the v blocks of the linked block design. Then the incidence matrix of supplemented design is

$$\mathbf{N}^* = \begin{bmatrix} \mathbf{N}_1' \\ \mathbf{I} \end{bmatrix}. \tag{7.6}$$

We then have the following theorem:

Theorem 7.8; The design with the incidence matrix (7.6) is SPEB with two efficiency classes and the parameters $v^* = b + v$, $b^* = v$, $\mathbf{r}^* = (k\mathbf{1}'\ \mathbf{1}')'$, $\mathbf{k}^* = (r+1)\mathbf{1}$, $\mu_1^* = (r + k - \lambda)/(rk + k)$, $\rho_1^* = v - 1$, $\mu_2^* = 0$, $\rho_2^* = b$, $\mathbf{L}_1^* = \{1/(r + k - \lambda)\}\left\{\begin{bmatrix} \mathbf{N}_1'\mathbf{N}_1 & \mathbf{N}_1' \\ k\mathbf{N}_1 & k\mathbf{I} \end{bmatrix} - k\,\mathbf{1}\,\mathbf{r}^{*\prime}/v\right\}$, $\mathbf{L}_2^* = \mathbf{I} - \mathbf{L}_1^* - \mathbf{1}\mathbf{r}^{*\prime}/n$ and $n = v(r+1)$.

The matrix $\mathbf{\Omega}$ for this design is

$$\mathbf{\Omega}^* = \begin{bmatrix} \mathbf{N}_1'\mathbf{N}_1/\lambda vk + \mathbf{I}/k - k\,\mathbf{1}\,\mathbf{1}'/\lambda v^2 & \mathbf{N}_1'/\lambda v - k\,\mathbf{1}\,\mathbf{1}'/\lambda v^2 \\ \mathbf{N}_1/\lambda v - k\,\mathbf{1}\,\mathbf{1}'/\lambda v^2 & (1 + k/\lambda v)\mathbf{I} - k\,\mathbf{1}\,\mathbf{1}'/\lambda v^2 \end{bmatrix}.$$

The adjusted treatment sum of squares is

$$(1 + k/\lambda v)\,\mathbf{Q}_2'\mathbf{Q}_2 + \mathbf{Q}_1'\mathbf{Q}_1/k + 2\mathbf{Q}_2'\mathbf{P}_1/\lambda v + \mathbf{P}_1'\mathbf{P}_1/\lambda vk,$$

where $\mathbf{P}_1 = \mathbf{N}_1'\mathbf{Q}_1$.

Hoblyn, Pearce and Freeman (1954) introduced supplemented balanced designs as a generalization of variance balanced designs. According to them a design is in supplemented balance if all treatments are replicated r times except the supplemented ones which have r_0 replications, and if all pairs of treatments concur λ times in blocks and the pairs having the supplemented treatment occur in λ_0 blocks. Pearce (1960) assumed that all blocks contain k plots. The treatments are numbered 0 to v, treatment 0 being the supplemented one. Let $\mathbf{N} = [\mathbf{n}_1\ \mathbf{n}_2 \ldots \mathbf{n}_b]$ be the treatments *vs* blocks incidence matrix having $(v+1)$ rows and b columns. Then Pearce defined the following

$$\lambda = \mathbf{n}_i'\mathbf{n}_j \quad \text{if neither } i \text{ nor } j \text{ equals } 0$$

$$\lambda_0 = \mathbf{n}_0'\mathbf{n}_j$$

$$s_j = \mathbf{n}_j'\mathbf{n}_j.$$

The matrix $\mathbf{N}\mathbf{N}'$ is of the form

$$\begin{bmatrix} s_0 & \lambda_0\mathbf{1}' \\ \lambda_0\mathbf{1} & (s - \lambda)\mathbf{I} + \lambda\,\mathbf{1}\,\mathbf{1}' \end{bmatrix}.$$

A class of designs similar to supplemented designs was discussed by Das (1958). These designs are called reinforced designs. A reinforced design is obtained by introducing $\alpha \geqslant 1$ new treatments all included in each of the blocks of an incomplete block design together with $\beta \geqslant 0$ new blocks each having $v + \alpha$ treatments in it. There are $v^* = v + \alpha$ treatments in $b^* = b + \beta$ blocks of size $\mathbf{k}^* = [(k + \alpha)\,\mathbf{1}'\ \ (v + \alpha)\,\mathbf{1}']'$ and replications $\mathbf{r}^* = [(r + \beta)\,\mathbf{1}'\,(b + \beta)\,\mathbf{1}']'$. It may be noted that when $\beta = 0$, we essentially have supplemented block designs.

7.4 DESIGNS WITH MERGED TREATMENTS

This Section is devoted to another useful class of designs obtained through merging the treatments in a given block design. Such designs can be obtained for v^* ($< v$) treatments from a block design in v treatments. The number of blocks and the block sizes of the resulting design remain the same as that of the original design. Moreover, the statistical properties of the original design are also retained in the resultant design.

The idea of collapsing (merging) the levels of factors was exploited by Addelman (1962) to construct orthogonal resolution–III plans for symmetrical and asymmetrical factorial experiments. The effect of merging of treatments in block designs was studied by Pearce (1971) and it was established that the merging of treatments in general leads to an increase in precision of the resulting design, if the merging has any effect. In many cases the merging of two treatments does not have any effect on the precision of the design obtained by merging the treatments. If two treatments are such that their interchanging has no effect on the matrix $\mathbf{\Omega}^{-1}$, then these treatments can be merged without really affecting the precision of the design. These treatments were termed as equivalent treatments by Pearce (1971). In randomized complete block and BIB designs all treatments are equivalent, while in supplemented block designs all treatments other than the supplemented treatments are equivalent. Similarly in GD design all the treatments within a group are equivalent. Puri and Nigam (1975b, 1983) have studied the effect of merging of treatments in EB, GD and factorially balanced (or designs having property A) designs. Before we present some of these results in detail, it is important to study first a Lemma due to Puri and Nigam (1978).

Consider a block design D (v, b, $\mathbf{r}$, $\mathbf{k}$). Let the v treatments be subdivided into an arbitrary number, say v^*, of disjoint groups such that the t-th group contains a_t treatments ($t = 1, \ldots, v^*$) so that $\Sigma a_t = v$. Suppose the a_t treatments of the t-th group are replaced by a single treatment, say A_t. This amounts to merging (adding) the elements of a_t rows (treatments in the t-th group) of the incidence matrix of D. Obtain the design D^* by merging the treatments of each of the t-groups. Clearly, the incidence matrix $\mathbf{N}$, the vector of replications $\mathbf{r}$, and the matrix $\mathbf{P}$ of the design D can be written in the partitioned form as

$$\mathbf{N} = \begin{bmatrix} \mathbf{N}_1 \\ \vdots \\ \mathbf{N}_{v^*} \end{bmatrix}, \mathbf{r} = \begin{bmatrix} \mathbf{r}_1 \\ \vdots \\ \mathbf{r}_{v^*} \end{bmatrix} \tag{7.7}$$

and

$$\mathbf{P} = ((\mathbf{P}_{tt'})),\ t, t' = 1, \ldots, v^*, \tag{7.8}$$

where $\mathbf{N}_t$ ($t = 1, \ldots, v^*$) is an $a_t \times b$ matrix, $\mathbf{r}_t$ is an a_t-component column vector and

$$\mathbf{P}_{tt'} = \mathbf{N}_t \mathbf{K}^{-1} \mathbf{N}_{t'}',\ t, t' = 1, \ldots, v^*, \tag{7.9}$$

is a partitioned matrix of order $a_t \times a_t'$. Then we prove the following Lemma for the design D^*:

Lemma 7.3: The elements $p^*_{tt'}$ of $\mathbf{P}^* = \mathbf{N}^*\mathbf{K}^{-1}\mathbf{N}^{*\prime}$ of the design D^* are the sum of the elements of the matrix $\mathbf{P}_{tt'} = \mathbf{N}_t\mathbf{K}^{-1}\mathbf{N}_{t'}$, $t, t' = 1, \ldots, v^*$, and the replication r_t of the t-th treatment in D^* is the sum of replications of all the a_t treatments in the t-th group of the design D.

Proof: The incidence matrix $\mathbf{N}^*$ of the design D^* is

$$\mathbf{N}^* = \begin{bmatrix} \mathbf{1}'\mathbf{N}_1 \\ \vdots \\ \mathbf{1}'\mathbf{N}_{v*} \end{bmatrix}.$$

The vector of replications $\mathbf{r}^*$ of the design D^* is

$$\mathbf{N}^*\mathbf{1} = (r_1^* \ldots r_{v*}^*) = (\mathbf{r}_1'\mathbf{1}\alpha \ldots \mathbf{r}_{v*}'\mathbf{1})'.$$

The proof is completed by noting that the element $p^*_{tt'}$ of the matrix $\mathbf{P}^*$ is

$$\begin{aligned} p^*_{tt'} &= \mathbf{1}'\,\mathbf{N}_t\mathbf{K}^{-1}\mathbf{N}_{t'}\,\mathbf{1} \\ &= \mathbf{1}'\mathbf{P}_{tt'}\mathbf{1}. \end{aligned}$$

7.4.1 Merging of treatments in EB designs

We now study the effect of merging of treatments in EB designs. We prove the following theorem:

Theorem 7.9: If D is an EB design, then D^* is also EB in $v^*\,(<v)$ treatments and with same efficiency factor as of D.

Proof: The $\mathbf{P}$ matrix of an EB design is

$$\mathbf{P} = \mu\mathbf{R} + (1-\mu)/n\mathbf{r}\mathbf{r}',$$

which can be written in the partitioned form as in (7.9) with partitioned matrices

$$\begin{aligned} \mathbf{P}_{tt} &= \mu\mathbf{R}_t + (1-\mu)/n\,\mathbf{r}_t\mathbf{r}_t',\ t = 1, \ldots, v^* \\ \mathbf{P}_{tt'} &= (1-\mu)/n\,\mathbf{r}_t\mathbf{r}_{t'}',\ t \neq t' = 1, \ldots, v^*, \end{aligned}$$

where $\mathbf{R}_t$ is a diagonal matrix with diagonal elements $\mathbf{r}_t$. Hence by Lemma 7.3, the elements of matrix $\mathbf{P}^*$ are

$$\begin{aligned} p^*_{tt} &= \mathbf{1}'\,[\mu\mathbf{R}_t + (1-\mu)/n\,\mathbf{r}_t\mathbf{r}_t']\,\mathbf{1} \\ &= \mu r_t^* + (1-\mu)r_t^{*\,2}/n \end{aligned}$$

and

$$p^*_{tt'} = (1-\mu)r_t^*\,r_{t'}^*.$$

Hence the matrix $\mathbf{P}^*$ of D^* is

$$\mathbf{P}^* = \mu\mathbf{R}^* + (1 - \mu)/n\ \mathbf{r}^*\mathbf{r}^{*\prime}$$

where $\mathbf{r}^*$ is the v^*-component vector of replications of D^* and $\mathbf{R}^*$ is a diagonal matrix with diagonal elements $\mathbf{r}^*$. Hence D^* is an EB design in v^* treatments with same efficiency factor μ. This completes the proof. For an alternate proof the readers may refer to Puri and Nigam (1975b).

It is known that an equi-replicate variance balanced design (v, b, r, k, ϕ) with

$$\mathbf{P} = \mathbf{D} + \phi\mathbf{1}\mathbf{1}'$$

is an EB design with efficiency factor $(r - v\,\phi)/r$.
We thus have the following:

Corollary 7.7: The design D^* obtained by merging of treatments in a variance balanced design $D\ (v, b, r, k, \phi)$ is EB with parameters

$$v^* \leqslant v,\ b^* = b,\ \mathbf{r}^* = r\mathbf{a},\ \mathbf{k}^* = k\mathbf{1},\ \mu^* = (r - v\phi)/r,$$
$$\mathbf{a} = (a_1, \ldots, a_{v*})'.$$

The design D^* is also variance balanced if $\mathbf{a} = a\mathbf{1}$.

It is also known that a BIB design is an EB design with $\mu = (r - \lambda)/rk$. We have the following:

Corollary 7.8: The design D^* obtained by merging treatments in a BIB design $D\ (v, b, r, k, \lambda)$ is an EB design with parameters

$$v^* \leqslant v,\ b^* = b,\ \mathbf{r}^* = r\mathbf{a},\ \mathbf{k}^* = k\ \mathbf{1},\ \mu^* = (r - \lambda)/rk.$$

We now illustrate the procedure through an example.

Example 7.1: Consider an EB design with parameters

$$v = 9,\ b = 11,\ \mathbf{r} = 5\ (1, 1, 1, 1, 1, 1, 1, 2, 2)',\ k = 5,\ \mu = 3/25$$

with block contents

(1, 3, 4, 5, 8) (2, 5, 7, 8, 8) (1, 2, 6, 8, 9)

(2, 4, 5, 6, 9) (3, 6, 8, 8, 9) (1, 2, 3, 7, 9)

(3, 5, 6, 7, 9) (4, 7, 8, 9, 9) (2, 3, 4, 8, 9)

(1, 4, 6, 7, 8) (1, 5, 8, 9, 9)

Let us merge the treatments (1, 2, 3) and (4, 5) through the following correspondence:

$$1 \to 1,\ 2 \to 1,\ 3 \to 1,\ 4 \to 2,\ 5 \to 2,\ 6 \to 3,\ 7 \to 4,$$
$$8 \to 5,\ 9 \to 6.$$

We thus get an EB design with parameters

$$v = 6,\ b = 11,\ \mathbf{r} = 5(3, 2, 1, 1, 2, 2)',\ k = 5,\ \mu = 3/25$$

and block contents

(1, 1, 2, 2, 5) (1, 2, 4, 5, 5) (1, 1, 3, 5, 6)
(1, 2, 2, 3, 6) (1, 3, 5, 5, 6) (1, 1, 1, 4, 6)
(1, 2, 3, 4, 6) (2, 4, 5, 6, 6) (1, 1, 2, 5, 6)
(1, 2, 3, 4, 5) (1, 2, 5, 6, 6)

Remark 7.1: In case of a variance balanced design D with parameters $v = av^*$, b, $\mathbf{r}$, $\mathbf{k}$, ϕ and with $\mathbf{NN}' = \mathbf{D} + \phi\mathbf{1}\,\mathbf{1}'$, merging of treatments arbitrarily does not yield an EB or variance balanced design. However, if the treatments are merged in groups of exactly 'a' treatments, then D^* will be variance balanced, but not EB, with parameters

$$v^* = v/a,\ b^* = b,\ r_t^* = \mathbf{1}'\mathbf{r}_t,\ t = 1,\ldots,v^*,\ \mathbf{k}^* = \mathbf{k},$$
$$\phi^* = a^2\phi,$$

and in D^* every elementary contrast is estimated with precision

$$2\sigma^2/(av\,\phi) < 2\sigma^2/(v\phi).$$

7.4.2 Merging of treatments in GD designs

We now consider the designs obtained by merging of treatments in GD designs. Consider a GD design D with parameters

$$v = m_1m_2,\ b,\ r,\ k,\ \lambda_1,\ \lambda_2$$

and let the treatments in D be arranged in m_1 groups of m_2 treatments each in the following array:

Group				
1	1	2	$\ldots j$	$\ldots m_2$
2	m_2+1	m_2+2	$\ldots m_2+j$	$\ldots 2m_2$
$\vdots$				
i	$(i-1)m_2+1$	$(i-1)m_2+2$	$\ldots(i-1)m_2+j$	$\ldots im_2$
$\vdots$				
m_1	$(m_1-1)m_2+1$	$(m_1-1)m_2+2$	$\ldots(m_1-1)m_2+j$	$\ldots m_1m_2.$

The treatments of each group are then divided into an arbitrary number of disjoint subgroups. The sub-group size of the i-th group is denoted by

$\mathbf{a}_i = (a_{i1},\ldots, a_{it_i})'$, such that $\sum_{j=1}^{t_i} a_{ij} = m_2$ and $1 \leqslant a_{ij} \leqslant m_2$, and $t_i \leqslant m_2$, $i = 1,\ldots, m_1$. The treatments of each sub-group are merged so as to get the design D^*. Clearly the design D^* has $v^* = \sum_{i=1}^{m_1} t_i$ treatments. The replication vector $\mathbf{r}^*$ of D^* is

$$\mathbf{r}^* = r(\mathbf{a}_1' \ldots \mathbf{a}_{m_1}')'.$$

We then have the following theorem:

Theorem 7.10: The design D^* is PEB with two efficiency classes and with parameters

$$\begin{aligned}
v^* &= \sum_{i=1}^{m_1} t_i,\ b^* = b,\ \mathbf{r}^* = r(\mathbf{a}_1' \ldots \mathbf{a}_{m_1}')',\\
\mathbf{k}^* &= k\mathbf{1},\ \mu_1^* = (r-\lambda_1)/rk,\ \mu_2^* = (rk - v\lambda_2)/rk,\\
\rho_1^* &= v^* - m_1,\ \rho_2^* = m_1 - 1, \qquad\qquad (7.10)\\
\mathbf{L}_1^* &= \mathbf{I}_{v^*} - \frac{1}{m_2}\,\mathrm{diag}\,(\mathbf{1a}_1', \ldots, \mathbf{1a}_{m_1}'),\\
\mathbf{L}_2^* &= \mathbf{I}_{v^*} - \mathbf{L}_0^* - \mathbf{L}_1^*,\ L_0^* = \mathbf{1}r^{*\prime}/rm_1m_2.
\end{aligned}$$

Proof: The matrix $\mathbf{P}$ of the design D can be expressed in the partitioned form as

$$\mathbf{P} = (\mathbf{P}_{ii'}),\ i, i' = 1,\ldots, m_1,$$

where $\mathbf{P}_{ii}$ and $\mathbf{P}_{ii'}$ are $m_2 \times m_2$ matrices given by

$$\begin{aligned}
\mathbf{P}_{ii} &= [(r-\lambda_1)\,\mathbf{I} + \lambda_1\mathbf{11}']/k,\ i = 1,\ldots, m_1,\\
\mathbf{P}_{ii'} &= \lambda_2\mathbf{11}'/k,\ \ i \neq i' = 1,\ldots, m_1.
\end{aligned}$$

The partitioned matrices $\mathbf{P}_{ii}$ and $\mathbf{P}_{ii'}$ can be further partitioned into submatrices corresponding to subgroups within a group and subgroups between groups in the following way:

$$\mathbf{P}_{ii} = (\mathbf{P}_{jj'}^{ii}),\ j, j' = 1, \ldots, t_i;\ i = 1,\ldots, m_1$$

and

$$\mathbf{P}_{ii'} = (\mathbf{P}_{jj'}^{ii'}),\ j = 1,\ldots, t_i;\ j' = 1, \ldots, t_{i'};$$
$$i \neq i' = 1,\ldots, m_1,$$

where the submatrices of order $a_{ij} \times a_{ij}$, $a_{ij} \times a_{ij'}$, and $a_{ij} \times a_{i'j'}$ are respectively

$$\begin{aligned}
\mathbf{P}_{jj}^{ii} &= [(r-\lambda_1)\,\mathbf{I} + \lambda_1\mathbf{11}']/k,\ j = 1,\ldots, t_i,\\
\mathbf{P}_{jj'}^{ii} &= \lambda_1\mathbf{11}'/k,\ j \neq j' = 1,\ldots, t_i
\end{aligned}$$

and

$$\mathbf{P}_{jj'}^{ii'} = \lambda_2\mathbf{11}'/k,\ j = 1,\ldots, t_i;\ j' = 1,\ldots, t_{i'};\ i \neq i'.$$

Using Lemma 7.3, the elements of $\mathbf{P}^*$ of the design D^* can be partitioned by adding all the elements of submatrices This gives matrix $\mathbf{P}^*$ as

$$\mathbf{P}^* = (\mathbf{P}^*_{ii'}),\ i, i' = 1,\ldots, m_1,$$

where the partitioned matrices of order $t_i \times t_i$ and $t_i \times t_{i'}$ are

$$\mathbf{P}^*_{ii} = [(r-\lambda_1)\mathbf{a}_i^{\delta} + \lambda_1 \mathbf{a}_i \mathbf{a}_i']/k,\ i = 1,\ldots, m_1$$

and

$$\mathbf{P}^*_{ii'} = \lambda_2 \mathbf{a}_i \mathbf{a}_{i'}'/k,\ i \neq i' = 1,\ldots, m_1.$$

Here $\mathbf{a}_i^{\delta}$ is a diagonal matrix with diagonal elements $\mathbf{a}_i$.

The matrix $\mathbf{M}_0^*$ of D^* can be written in partitioned form as

$$\mathbf{M}_0^* = (\mathbf{M}^0_{ii'}),\ i, i' = 1,\ldots, m_1,$$

where the partitioned matrices of order $t_i \times t_i$ and $t_i \times t_{i'}$ are

$$\mathbf{M}^0_{ii} = [v(r-\lambda_1)\,\mathbf{I} - (rk - v\lambda_1)\mathbf{1}\mathbf{a}_i']/rk\,v,\ i = 1,\ldots, m_1,$$

$$\mathbf{M}^0_{ii'} = -\,(rk - v\lambda_2)\mathbf{1}\mathbf{a}_{i'}'/rk\,v,\ i \neq i' = 1,\ldots, m_1.$$

It can be seen easily that the non-zero eigen values of the matrix $\mathbf{M}_0^*$ are $(r-\lambda_1)/rk$ and $(rk - v\lambda_2)/rk$ with multiplicities $v^* - m_1$ and $m_1 - 1$ respectively.

The $\sum_{i=1}^{m_1} (t_i - 1) = v^* - m_1$ contrasts, which are eigen vectors of $\mathbf{M}_0^*$ corresponding to eigen value μ_1^*, can be written as

$$\mathbf{s}^*_{1u} = \begin{bmatrix} \mathbf{s}_{1u} \\ \mathbf{0} \\ \vdots \\ \mathbf{0} \end{bmatrix},\quad \mathbf{s}_* = \begin{bmatrix} \mathbf{0} \\ \mathbf{s}_{2u} \\ \mathbf{0} \\ \vdots \\ \mathbf{0} \end{bmatrix},\ldots,\ \mathbf{s}^*_{m_1u} = \begin{bmatrix} \mathbf{0} \\ \mathbf{0} \\ \vdots \\ \mathbf{s}_{m_1u} \end{bmatrix},$$

where $\mathbf{0}$ denotes the null vector of appropriate dimension and $\mathbf{s}_{iu}$, $i=1,\ldots m_1$; $u = 1,\ldots, t_i - 1$ denotes the complete set of $t_i - 1$ mutually orthogonal contrasts among t_i treatments of i-th group satisfying $\mathbf{s}_{ij}'\mathbf{r}_i^* = 0$ and $\mathbf{s}_{ij}'\mathbf{r}_i^{*\delta}\mathbf{s}_{ij'} = 0$, $j \neq j', j, j' = 1,\ldots, t_i - 1$. Here $\mathbf{r}_i^{*\delta}$ is a diagonal matrix with diagonal elements $\mathbf{r}_i^*$.

If $G_i (i = 1,\ldots, m_1)$ denotes the total yield of all the t_i treatments of i-th group based on rm_2 observations, then it can be shown that all the $(m_1 - 1)$ contrasts between G_i totals are eigen vectors of $\mathbf{M}_0^*$ corresponding to eigen value μ_2^*. Hence all these contrasts are basic contrasts. The $\mathbf{L}_1^*$ and $\mathbf{L}_2^*$ matrices as given in (7.10) are easy to obtain. Hence the theorem.

Corollary 7.9: If D is singular or semi regular then the design D^* is SPEB.

Corollary 7.10: If $\lambda_1 = 0$, then the design D^* is proper, binary PEB with varying replications.

If in addition to $\lambda_1 = 0$ in Corollary 7.10, we have $m_2 = at$, then by merging 'a' treatments in each of t disjoint sets or subgroups, we have the following:

Corollary 7.11: The existence of GD design D with parameters $v = m_1 m_2$, $b, r, k, \lambda_1 = 0, \lambda_2, m_2 = at$ implies the existence of a GD design D^* with parameters

$$v^* = m_1 t,\ b^* = b,\ r^* = ar,\ k^* = k,\ \lambda_1^* = 0,\ \lambda_2^* = a^2\lambda_2.$$

Designs based on Corollary 7.11 have been constructed by Dey and Nigam (1985). Through this procedure several new GD designs of regular type were reported by them.

Example 7.2: Consider a semi regular GD design with parameters $v = 12$, $b = 9$, $r = 3$, $k = 4$, $m_1 = 4$, $m_2 = 3$, $\lambda_1 = 0$, $\lambda_2 = 1$, and block contents

(1, 4, 7, 10) (1, 5, 8, 11) (1, 6, 9, 12) (2, 6, 8, 10) (2, 4, 9, 11)

(3, 4, 8, 12) (3, 5, 9, 10) (3, 6, 7, 11) (2, 5, 7, 12)

Here the treatment groups are (1, 2, 3) (4, 5, 6) (7, 8, 9), (10, 11, 12). Obtain D^* by merging treatment numbers (8, 9) and (10, 11) respectively. The design D^* so obtained will be proper and binary with parameters $v^* = 10$, $b^* = 9$, $\mathbf{r}^* = 3(1, 1, 1, 1, 1, 1, 1, 2, 2, 1)'$, $k^* = 4$ and block contents

(1, 4, 7, 9) (1, 5, 8, 9) (1, 6, 8, 10) (2, 6, 8, 9) (2, 4, 8, 9)

(3, 4, 8, 10) (3, 5, 8, 9) (3, 6, 7, 9) (2, 5, 7, 10)

and D^* is SPEB with $\mu = 1/4$ and

$$\mathbf{L}^* = \begin{bmatrix} \mathbf{A} & \mathbf{0} & \mathbf{0} & \mathbf{0} \\ \mathbf{0} & \mathbf{A} & \mathbf{0} & \mathbf{0} \\ \mathbf{0} & \mathbf{0} & \mathbf{B} & \mathbf{0} \\ \mathbf{0} & \mathbf{0} & \mathbf{0} & \mathbf{C} \end{bmatrix}$$

where

$$\mathbf{A} = \begin{bmatrix} 2 & -1 & -1 \\ -1 & 2 & -1 \\ -1 & -1 & 2 \end{bmatrix},\ \mathbf{B} = \begin{bmatrix} 2 & -2 \\ -1 & 1 \end{bmatrix},\ \mathbf{C} = \begin{bmatrix} 1 & -1 \\ -2 & 2 \end{bmatrix}$$

7.5 DESIGNS WITH MISSING BLOCKS

This Section deals with the problem of analysis of incomplete block designs when observations corresponding to all the treatments in a block are missing. Such a situation arises in experiments in physical and animal sciences where the blocks are natural units. For instance, an incubator used in a hatching experiment of chicken as a block may be out of order at the time of experimentation. Similarly a particular litter, acting as a block, may not be available at the time of experimentation. It will be seen later that the analysis of such designs does not pose any new problems and these designs can be analysed as PEB designs.

The earliest work on this problem is from Cornish (1940) who gave the analysis of BIB designs with one missing block. Zelen (1954) and Puri (1984) extended this to the case of two associate PBIB designs. We present here the work based on Puri (1984).

7.5.1 Missing block in BIB design

Consider a BIB design $D\ (v, b, r, k, \lambda)$ with v treatments denoted by $t_1, \ldots, t_v$. Suppose all the k treatments of a block are missing. Without any loss of generality, let the missing block contain the treatments $t_1, \ldots, t_k$. Call the resultant design as D^*. Clearly D^* has v treatments, $(b-1)$ blocks each of size k, the affected treatments have $(r-1)$ replications, while the unaffected treatments have r replications. Any pair of treatments which is affected appears in $\lambda - 1$ blocks, whereas all other pairs of treatments appear in λ blocks. The concurrence matrix of the design D^* can be written as

$$\mathbf{N}^*\mathbf{N}^{*\prime} = \begin{bmatrix} \mathbf{A} & \mathbf{B} \\ \mathbf{B}' & \mathbf{C} \end{bmatrix} \tag{7.11}$$

with partitioned matrices as

$$\mathbf{A} = (r-\lambda)\,\mathbf{I}_k + (\lambda - 1)\,\mathbf{1}_k\mathbf{1}_k',$$

$$\mathbf{B} = \lambda\mathbf{1}_k\mathbf{1}_{v-k}'$$

$$\mathbf{C} = (r-\lambda)\,\mathbf{I}_{v-k} + \lambda\mathbf{1}\,\mathbf{1}'.$$

The vector of replications $\mathbf{r}^*$ of D^* is

$$\mathbf{r}^* = [(r-1)\,\mathbf{1}_k' \quad r\mathbf{1}_{v-k}']' \tag{7.12}$$

and the matrix $\mathbf{M}$ of D^* is

$$\mathbf{M}^* = \{1/rk(r-1)\} \begin{bmatrix} r\mathbf{A} & r\mathbf{B} \\ (r-1)\mathbf{B}' & (r-1)\mathbf{C} \end{bmatrix}. \tag{7.13}$$

The pattern of matrices $\mathbf{N}^*\mathbf{N}^{*\prime}$ and $\mathbf{M}^*$ are similar to those of supplemented balanced designs of Pearce (1960). It can be shown easily that the eigen values of the matrix $\mathbf{M}^*$ other than 1 are

$$\mu_1 = (r - \lambda)/r(k - 1),\ \mu_2 = (r - \lambda)/rk,\ \mu_3 = (r - \lambda)\ (r - k)/rk(r - 1) \tag{7.14}$$

with multiplicities $(k - 1)$, $(v-k-1)$ and 1, respectively. In other words D^* is a PEB design with three efficiency classes.

Let $\mathbf{s}_u$, $u = 1,\ldots, k - 1$, be the $k - 1$ mutually orthogonal contrasts among k affected treatments and $\mathbf{s}_u^*$, $u = k, k + 1,\ldots, v - 2$ be the $v - k - 1$ mutually orthogonal contrasts among unaffected treatments. Then the complete set of $v - 1$ mutually orthogonal contrasts can be written as

$$\mathbf{s}_u^{(1)} = (\mathbf{s}_u'\ \mathbf{0}_{v-k}')',\ u = 1,\ldots, k - 1, \tag{7.15}$$

$$\mathbf{s}_u^{(2)} = (\mathbf{0}_k'\ \mathbf{s}_u^{*\prime})',\ u = k,\ k + 1,\ldots, v - 2, \tag{7.16}$$

$$\mathbf{s}_u^{(3)} = [(v - k)\ r\mathbf{1}_k'\ (1 - r)\ k\mathbf{1}_{v-k}']',\ u = v - 1. \tag{7.17}$$

It is seen that these contrasts are eigen vectors of the matrix $\mathbf{M}^*$ corresponding to eigen values μ_1, μ_2 and μ_3 respectively, and, therefore, are basic contrasts. It may be noted that the last intra-group contrast between affected and unaffected treatments is estimated with more efficiency in D^* as compared to D because $\mu_3 < (r - \lambda)/rk$. In case the BIB design is symmetric, $\mu_3 = 0$.

Example 7.3: Consider a BIB design $v=b=7, r = k = 3, \lambda = 1$ with block contents (1, 2, 4), (2, 3, 5), (3, 4, 6), (4, 5, 7), (5, 6, 1), (6, 7, 2), (7, 1, 3). Let the first block containing treatments 1, 2 and 4 be missing. Let us renumber the treatments as explained earlier, i.e. 3↔4. Then the design D^* is PEB with three efficiency classes and parameters

$$v = 7,\ b = 6,\ \mathbf{r}^* = (2, 2, 2, 3, 3, 3, 3)',\ k = 3,$$

$$\mu_1 = 1/3,\ \mu_2 = 2/9,\ \mu_3 = 0,\ \rho_1 = 2,\ \rho_2 = 3,\ \rho_3 = 1.$$

7.5.2 Missing block in semi regular GD design

Bose and Connor (1952) have shown that for semi regular GD designs $k = cm$, where c is some integer, and each block contains exactly c treatments from each group. Here m denotes the total number of groups. Consider an $m_1 \times m_2$ factorial experiment laid out in a semi regular GD design with parameters $v = m_1 m_2$ b, r, $k = cm_1$, λ_1, λ_2. Let $i = (i_1, i_2)$ denote the level combination having first factor at i_1-th level and the second factor at i_2-th level, $i_j = 0, 1,\ldots, m_j - 1$; $j = 1, 2$. Let all the k observations of one block be missing. Without any loss of generality, let the contents of the missing block be

$$(i_1, 1),\ \ldots,\ (i_1, c),\ i_1 = 0, 1,\ldots, m_1 - 1.$$

Let the resultant design be D^*. If we arrange the treatments of the i_1-th group as

$$(i_1, 1), \ldots, (i_1, c), (i_1, c+1), \ldots, (i_1, m_2)$$

and write the treatments in such a fashion that the first m_2 belong to first group, the next m_2 treatments belong to second group, and so on, and the last m_2 treatments belong to the m_1-th group, then the concurrence matrix of the design D^* can be written as

$$\mathbf{N^*N^{*\prime}} = \{\mathbf{A}, \mathbf{B}, \ldots, \mathbf{B}\}.$$

The partitioned matrices $\mathbf{A}$ and $\mathbf{B}$ of dimension $m_2 \times m_2$ can be further partitioned as

$$\mathbf{A} = \begin{bmatrix} \mathbf{A}_{11} & \mathbf{A}_{12} \\ \mathbf{A}_{12}' & \mathbf{A}_{22} \end{bmatrix} \text{ and } \mathbf{B} = \begin{bmatrix} \mathbf{B}_{11} & \mathbf{B}_{12} \\ \mathbf{B}_{12}' & \mathbf{B}_{22} \end{bmatrix}$$

where

$$\mathbf{A}_{11} = (r - \lambda_1)\, \mathbf{I}_c + (\lambda_1 - 1)\, \mathbf{1}\,\mathbf{1}', \ \mathbf{A}_{12} = \lambda_1\, \mathbf{1}_c\, \mathbf{1}_{m_2-c}',$$

$$\mathbf{A}_{22} = (r - \lambda_1)\, \mathbf{I}_{m2-c} + \lambda_1 \mathbf{1}\,\mathbf{1}', \ \mathbf{B}_{11} = (\lambda_2 - 1)\, \mathbf{1}_c \mathbf{1}_c',$$

$$\mathbf{B}_{12} = \lambda_2 \mathbf{1}_c\, \mathbf{1}_{m_2-c}', \ \mathbf{B}_{22} = \lambda_2 \mathbf{1}_{m_2-c}\, \mathbf{1}_{m_2-c}'.$$

The replication vector $\mathbf{r}^*$ of D^* is

$$\mathbf{r}^* = \mathbf{1}_{m_1} \otimes \mathbf{r}_{m_2}, \mathbf{r}_{m_2} = [(r-1)\, \mathbf{1}_c' \ \ r\mathbf{1}_{m_2}' \ \ c]'.$$

The matrix $\mathbf{M}$ of D^* is

$$\mathbf{M}^* = [\mathbf{I}_{m_1} \otimes (\mathbf{A}^* - \mathbf{B}^*) + \mathbf{1}_{m_1} \mathbf{1}_{m_1}' \otimes \mathbf{B}]/k,$$

where $\mathbf{A}^* = \mathbf{R}_{m_2}^{-1}\, \mathbf{A}$ and $\mathbf{B}^* = \mathbf{R}_{m_2}^{-1}\, \mathbf{B}$ and $\mathbf{R}_{m_2}$ is a diagonal matrix with diagonal elements as $\mathbf{r}_{m_2}$.

It can be seen easily that the eigen value of $\mathbf{M}^*$ other than 1 are

$$\mu_1 = 0, \ \mu_2 = (r - \lambda_1)/(r-1)\, k, \ \mu_3 = (r - \lambda_1)/rk,$$

$$\mu_4 = [(r - \lambda_1)(r - c) - (k - c)(r - \lambda_2)]/rk\,(r-1),$$

$$\mu_5 = [(r - \lambda_1)\, r - c\,(\lambda_2 - \lambda_1)]/rk\,(r-1),$$

with multiplicities $(m_1 - 1)$, $m_1(c-1)$, $m_2(m_2 - c - 1)$, 1 and $(m_1 - 1)$ respectively. Hence the design D^* is a PEB with five (or less) efficiency classes. It will have atmost four efficiency classes if $c = 1$ or

$$\lambda_1 = \lambda_2\, [1 - (v-k)/(rm_2 - c)] \quad \text{or} \quad m_2 = c + 1.$$

Let $\mathbf{s}_{1u}$, $u = 1, \ldots, m_1 - 1$ represent the complete set of $(m_1 - 1)$ mutually orthogonal contrasts associated with the m_1 levels of first factor and $\mathbf{s}_{2u}$ (c-component vector, $u = 1, \ldots, c-1$) and $\mathbf{s}_{2u}^*$ ($m_2 - c$ component vector, $u = c, c+1, \ldots, m_2 - 2$) denote the $c - 1$ and $m_2 - c - 1$ mutually orthogonal treatment contrasts among c affected and $(m_2 - c)$ unaffected levels of second factor, respectively. Then the $(m_2 - 1)$ mutually orthogonal contrasts associated with the m_2 levels of second factor can be written as

$$\mathbf{s}_{2u}^{(1)} = (\mathbf{s}_{2u}', \mathbf{0}')', \; u = 1, \ldots, c - 1,$$

$$\mathbf{s}_{2u}^{(2)} = (\mathbf{0}', \mathbf{s}_{2u}^{*\prime})', \; u = c + 1, \ldots, m_2 - 2,$$

$$\mathbf{s}_{2u}^{(3)} = [(m_2 - c)\, r\, \mathbf{1}_c' \; (1 - r)\, c\, \mathbf{1}_{m_2 - c}']', \; u = m_2 - 1.$$

The complete set of mutually orthogonal contrasts corresponding to the effect $A_1^{X_1} A_2^{X_2}$, $X_1 = 0$ or 1 and $X_2 = 0, 1, 2, 3$ can be written as

$$\mathbf{s}_{1u}^{X_1} \otimes \mathbf{s}_{2u}^{X_2}, \; u = 1, \ldots, m_j - 1; j = 1, 2,$$

where X_1 and X_2 cannot be zero simultaneously, where

$$\begin{aligned} \mathbf{s}_{2u}^{X_2} &= \mathbf{s}_{2u} \quad \text{if } X_2 = 1, 2, 3, \\ &= \mathbf{1}_{m_2} \text{ otherwise,} \end{aligned}$$

and

$$\begin{aligned} \mathbf{s}_{1u}^{X_1} &= \mathbf{s}_{1u} \quad \text{if } X_1 = 1 \\ &= \mathbf{1}_{m_1} \text{ otherwise.} \end{aligned}$$

It is easy to verify that all these contrasts are basic contrasts. The $(m_1 - 1)$ contrasts corresponding to main effect A_1 with $X_1 = 1$, $X_2 = 0$ are eigen vectors of $\mathbf{M}^*$ corresponding to eigen value 0, and the $(c - 1)$, $(m_2 - c - 1)$ and 1 contrasts with $X_1 = 0$, $X_2 = 1, 2, 3$ among affected levels, unaffected levels and inter group for main effect A_2 are eigen vectors of $\mathbf{M}^*$ corresponding to the eigen values μ_2, μ_3 and μ_4 respectively, and the contrasts corresponding to interaction A_1A_2 with $X_1 = 1$, $X_2 = 1, 2, 3$ are estimated with losses μ_2, μ_3 and μ_4, respectively.

Remark 7.2: It is clear that the design D^* has factorial structure. It is factorially balanced if $m_2 = 2$ and $c = 1$.

Example 7.4: Consider the semi regular GD design with parameters $v = 14$, $b = 8$, $r = 4$, $k = 7$, $\lambda_1 = 0$, $\lambda_2 = 3$, $m_1 = 7$, $m_2 = 2$ and $c = 1$ with block contents

(1, 2, 3, 4, 5, 6, 7), (8, 9, 10, 11, 5, 6, 7), (1, 2, 10, 11, 12, 13, 7),
(8, 9, 3, 4, 12, 13, 7), (1, 9, 3, 11, 5, 13, 14), (8, 2, 10, 4, 5, 13, 14),
(1, 9, 10, 4, 12, 6, 14), (8, 2, 3, 11, 12, 6, 14).

Let all the treatments of first block be missing. The design is SPEB with non zero eigen value as 1/6.

The new design is not only factorially balanced but is also more efficient, needs less number of points and provides the estimates of both the main effects A_1 and A_2 orthogonal to block effects.

7.5.3.1 Missing block containing all the treatments of any c groups

Consider again a two factor factorial experiment $m_1 \times m_2$ in GD design with parameters

$$v = m_1 m_2,\ b,\ r,\ k = cm_2,\ \lambda_1,\ \lambda_2.$$

Without any loss of generality let all the k treatments of first c groups be missing. If we arrange the treatments as described in Section 7.5.2., then the concurrence matrix of D^* can be written in partitioned form as in (7.11), where the partitioned matrices are

$$\mathbf{A} = (r - \lambda_1)\, \mathbf{I}_c \otimes \mathbf{I}_{m_2} + [(\lambda_1 - \lambda_2)\, \mathbf{I}_c + (\lambda_2 - 1)\mathbf{1}\,\mathbf{1}'] \otimes \mathbf{1}_{m_2}\mathbf{1}'_{m_2}$$

$$\mathbf{B} = \lambda_2\, \mathbf{1}_c \mathbf{1}'_{m_1 - c} \otimes \mathbf{1}_{m_2}\mathbf{1}'_{m_2}$$

$$\mathbf{C} = (r - \lambda_1)\, \mathbf{I}_{m_1 - c} \otimes \mathbf{I}_{m_2} + [(\lambda_1 - \lambda_2)\, \mathbf{I}_{m_2 - c} + \lambda_2 \mathbf{1}\,\mathbf{1}'] \otimes \mathbf{1}_{m_2}\mathbf{1}'_{m_2}.$$

The vector of replications $\mathbf{r}^*$ of D^* is

$$\mathbf{r}^* = \mathbf{r}_{m_1} \otimes \mathbf{1}_{m_2}$$

with $$\mathbf{r}_{m_1} = [(r - 1)\, \mathbf{1}'_c \;\; r\mathbf{1}'_{m_1 - c}]'$$

and the matrix $\mathbf{M}$ of D^* is as given in (7.13). It can be seen that the design D^* is PEB with five efficiency classes, the losses being respectively,

$$\mu_1 = (rk - v\lambda_2)/(r - 1)k,\ \mu_2 = (rk - v\lambda_2)/rk,$$

$$\mu_3 = (rk - v\lambda_2)/rk - (v - k)\lambda_2/r\,(r - 1)k,$$

$$\mu_4 = (r - \lambda_1)/(r - 1)k \text{ and } \mu_5 = (r - \lambda_1)/rk,$$

with multiplicities $(c - 1)$, $(m_1 - c - 1)$, 1, $(m_2 - 1)c$, $(m_1 - c)\,(m_2 - 1)$, respectively.

When D is singular, the design D^* is PEB with four efficiency classes. It may have three efficiency classes if $m_1 = c + 1$ or $k = m_2\, r$. The contrasts corresponding to main effect A_1 are eigen vectors of $\mathbf{M}^*$ corresponding to eigen values μ_1, μ_2 and μ_3 with multiplicities $c - 1$, $m_1 - c - 1$, and 1 respectively and contrasts corresponding to main effect A_2 and interaction A_1A_2 are eigen vectors corresponding to eigen value 0.

7.6 α-DESIGNS

This Section is devoted to another useful class of designs, called α designs, introduced by Patterson and Williams (1976a). These are resolvable incomplete block designs and are available for a wide variety of parametric combinations. The concept of resolvability has been introduced in Chapter 2. We have noticed that besides the well known class of BIB, Lattice and PBIB designs, others like cyclic designs can also be resolvable. As already stated in Chapter 6, the property of resolvability can be used to eliminate an additional source of variation.

For instance, field trials involving a large number of crop varieties not only require the use of incomplete block designs for controlling within block error but also need to be resolvable so that it will be possible to lay out the trials in more than one season (one replication per season) so that the effects due to seasons could be taken care of.

A limitation with the conventional resolvable designs is that these are available for limited combinations of the number of treatments v and the block size k. Resolvable square lattice designs are available for the combinations $v = s^2$ and $k = s$, where s is the number of blocks per replication, with further restrictions on v for higher order lattices (Yates, 1949). Again resolvable simple and triple rectangular lattices are available for $v = s(s-l)$ and $k = s-l$, where $s > l$ (Kempthorne, 1952). Resolvable cyclic designs (David, 1967) also have the restriction that either $k = r$ or is a multiple of r. The α designs introduced by Patterson and Williams have no restriction on the block size except that the number of treatments should be of the form ks or $k_1s_1 + k_2s_2$ (for the cases of equi-block or two different block sized designs respectively) in the acceptable range of block sizes k, k_1 and k_2.

Let $\boldsymbol{\alpha}$ be a $k \times r$ array with elements $a(p, q)$, $p = 1, \ldots, k$; $q = 1, \ldots, t$, in the set of residues mod s. Let $\boldsymbol{\alpha}^*$ be the array of size $k \times rs$ obtained by generating $s - 1$ further columns through cyclic substitutions from each column of $\boldsymbol{\alpha}$. If we add s to all elements in row 2, $2s$ to all elements in row 3, and so on, and $(k-1)s$ to all elements in row k of $\boldsymbol{\alpha}^*$, then the resulting array has symbols $0, 1, \ldots, v-1$. These symbols represent the v treatments (or varieties). The resulting $k \times rs$ array is the required α-design with columns as blocks and each set of s columns generated from the same column of $\boldsymbol{\alpha}$ constitutes a complete replication. The parameters of the design are $v = ks$, $b = rs$, r, k.

The array used could be of a particular kind, called reduced array, with all its first row and first column elements as zero, since designs obtained from other arrays are isomorphic to the one from this reduced array. The designs are isomorphic because the reduced array can be obtained from any given array by the rearrangement of blocks and by relabelling of treatments. Other operations generating isomorphic designs include permutation of row and/or columns and multiplication of all elements of the array with an integer which is coprime to s. The number of times any two treatments, say i and j, occur together in a block can be determined merely by inspecting $\boldsymbol{\alpha}$ without constructing the complete design. This is given by the frequency of $(j-i)$ mod s, where $j > i$, in the set of r differences $\{a(p_j, q) - a(p_i, q)\}$ mod s, where p_i is one more than the integral part of i/s. The design with the concurrences $\lambda_1, \lambda_2 \ldots$, is called an $\alpha(\lambda_1, \lambda_2, \ldots)$ design. It is seen that the integral parts of i/s and j/s can not be equal. Hence, neither balanced designs nor $\alpha(1, 2)$ designs can be constructed. Patterson and Williams (1976a) suggest a computer algorithm for the construction of α-designs which are optimal with regard to some criterion.

α-designs with unequal block sizes can also be constructed. Let $v = s_1k_1 + s_2k_2$, where s_1, s_2, k_1, k_2 are positive integers and k_1 and k_2 $(k_1 > k_2)$ are acceptable block sizes. We construct an α-design with $v' = v + (k_1 - k_2)s_2$ treatments with $s = s_1 + s_2$ blocks of k_1 plot each per replication, since $v' = k_1(s_1 + s_2)$, and delete $(k_1 - k_2)$ sets of s_2 varieties no two of which occur together in a block. The varieties numbered $v, v + 1, \ldots, v + s_2 - 1$ form such a set when $k_2 = k_1 - 1$.

The main advantage of α-designs is that these are not only resolvable but are available for all situations with $v = ks$ or $v = k_1s_1 + k_2s_2$. Suitably chosen α(0, 1) or α(0, 1, 2) designs can serve as substitutes for lattice designs with little loss of efficiency. It is known that rectangular Lattice with the parameters $v = 30$, $k = 5$, $r = 4$, and square lattice with the parameters $v = 36$, $k = 6$, $r = 4$, with distinct blocks do not exist.* However, α-designs exist for both these parametric combinations.

Example 7.5: Let $s = 3$, $r = k = 2$. Consider the α-array as given below:

$$\begin{matrix} 0 & 0 \\ 0 & 1 \end{matrix}$$

Each column of α generates two additional columns by cyclically adding 1 to each element of the column and reducing mod 3. This gives the 2×6 array α* as given below

$$\begin{matrix} 0 & 1 & 2 & 0 & 1 & 2 \\ 0 & 1 & 2 & 1 & 2 & 0 \end{matrix}$$

Now the required α design is obtained by adding 3 to the elements of the second row of α*. The design is

$$\begin{matrix} 0 & 1 & 2 & 0 & 1 & 2 \\ 3 & 4 & 5 & 4 & 5 & 3 \end{matrix}$$

The design is α(0, 1) and has other parameters as $v = b = 6$, $r = k = 2$.

The α-designs with $k = 2$ are useful as paired comparison designs. Paired comparison designs are incomplete block designs with two units per block. David (1967) discussed the use of these designs for tournaments and considered the construction of resolvable paired comparison designs. Williams (1976) discussed the class of resolvable paired comparison designs and

*The square lattice design $v = 36$, $k = 6$, $r = 4$ can be obtained by duplicating the design $v = 36$, $k = 6$, $r = 2$.

derived their canonical efficiency factors (James and Wilkinson, 1971). Comparisons are made between these and cyclic paired comparison designs in terms of the efficiency factors and it is observed that α-designs are as good as cyclic designs in terms of efficiency factors but have the advantage of being resolvable. For details on these and on some further work on the efficiency and optimality of two-replicate resolvable designs the interested reader is advised to refer to Williams, Patterson and John (1976).

7.7 NEARLY BALANCED DESIGNS

In Chapter 3 we have studied some well known balanced and partially balanced designs. Balanced designs are useful because they provide the estimates of all elementary contrasts with same precision. Balanced designs have been found to be very fascinating both by the theoreticians and the experimenters because of the ease of their analysis. However, for a given parametric combination (v, b, $\mathbf{r}$, $\mathbf{k}$), a balanced design may be non-existent. In such a situation one is forced to adopt a design which is as balanced as possible. Nigam (1976) defined nearly balanced designs to meet such requirements. A design D (v, b, $\mathbf{r}$, $\mathbf{k}$) is called nearly balanced if variance of all the elementary contrasts is nearly same. Such designs have an advantage that they can be laid out with fewer number of experimental units than balanced designs. Further, if high degree of accuracy is not desired, then the analysis of nearly balanced designs is fairly simple. For an interesting reading, reference may be made to Nigam (1976) and Verdooren (1966).

Cheng and Wu (1981) also defined nearly balanced designs. Their definition, however, is restricted to binary and proper designs. While the main thrust is to choose designs which are as balanced as possible, the emphasis is more on combinatorial balance instead of variance balance. Utilizing the idea of John and Mitchell (1977) that the two associate PBIB designs with $\lambda_2 = \lambda_1$ or $\lambda_2 = \lambda_1 + 1$ turn out to be very efficient, Cheng and Wu extended this idea to include designs with varying replications.

The motivation to develop such designs originates from the fact that although BIB designs are universally optimal (see Chapter 8) these may not exist for arbitrary v, b, k. The best alternative may then be to choose a design whose combinatorial properties may be as close to those of BIB design as possible. Cheng and Wu defined a nearly balanced design as follows:

Definition 7.4: An incomplete block design in v treatments, b blocks each of size k is called a nearly balanced design if it satisfies the following conditions:

i. Each treatment appears in each block at most once.
ii) Each $r_i = \bar{r}$ or $\bar{r} + 1$, where r_i is the replication of i-th treatment, $i = 1, \ldots, v$, and $\bar{r} = [bk/v]$ is the integral part of bk/v.

iii) For each fixed i_0, with $j, j' \neq i_0, j \neq j'$,

$$|\lambda_{i_0 j} - \lambda_{i_0 j'}| \leqslant 1,$$

where λ_{ij} is the number of times treatments i and j occur together.

Cheng showed that the existence of a nearly balanced incomplete block design implies the existence of a graph with atmost two distinct degrees which can be determined from the values of v, b and k. Utilizing this, necessary conditions can be derived for the existence of nearly balanced designs. For more details, reader may refer to Cheng and Wu (1981).

These designs can be further classified into two types. Let s be the number of treatments i with $r_i = \bar{r}$. Then $s = v - (bk - v\bar{r})$, $1 \leqslant s \leqslant v$. For $s = v$, the design is a regular graph design. From the definition, it is immediate that if $r_{i0} = \bar{r}$, then treatment i_0 appears together with any other treatment $\bar{\lambda}$ or $\bar{\lambda}+1$ times, where $\bar{\lambda}$ is the integral part of $(k-1)\,\bar{r}/(v-1)$. In this case, the number of treatments each of which appears together with treatment i_0 just $\bar{\lambda}$ times is $n = v-1-\{(k-1)\,\bar{r} - \bar{\lambda}(v-1)\}$. As to the treatments i_0 with $r_{i0} = \bar{r}+1$, there are two possibilities according as $n \geqslant k-1$ or $n < k-1$. If $n \geqslant k-1$, then, for $r_{i_0} = \bar{r}+1$, there are $n-k+1$ treatments each of which appears together with treatment i_0 just $\bar{\lambda}$ times, and there are $v+k-n-2$ $\lambda_{i_0 j}$'s equal to $\bar{\lambda}+1$. On the other hand, if $n < k-1$, then, for $r_{i_0} = \bar{r}+1$, there are $v-k+n$ treatments each of which appears together with treatment i_0 just $\bar{\lambda}+1$ times, and there are $k-1-n$ $\lambda_{i_0 j}$'s equal to $\bar{\lambda}+2$. Thus if $n \geqslant k-1$, then there are only two possible values of λ_{ij}, that is $\bar{\lambda}$ or $\bar{\lambda}+1$. Such a nearly balanced incomplete block design is said to be of type I. For $n < k-1$, λ_{ij} can be $\bar{\lambda}$, $\bar{\lambda}+1$, or $\bar{\lambda}+2$. In this case we have a nearly balanced incomplete block design of type II.

7.8 TREND FREE DESIGNS

Throughout this book we have been concentrating our discussions on designs useful for one-way elimination of heterogeneity. The entire experimental material is divided into homogeneous blocks and the treatments are assigned randomly to the units within blocks. Many experimental situations arise in which the response may be affected by the spatial or temporal position of the experimental unit or plot within a block and, frequently, an assumption of a common polynomial trend of specified degree over plots within blocks may be appropriate. A $v \times v$ latin square may be useful for eliminating the effects of a common, one dimensional, polynomial trend to degree $v-1$ over the plots in rows of the square. The criteria for and the optimality of other block designs, complete or incomplete, that eliminate the effects of common polynomial trends of specified degree over the plots within blocks have been investigated by Bradley and Yeh (1980) and Yeh and Bradley (1983). Consider a binary block design $D(v,b,k)$. A common polynomial trend is assumed to exist over the plots in each block. The trend may be in one or several dimensions and is expressed in terms of $p \leqslant (k-1)$ orthogonal components. In the usual intra block model for the general block designs orthogonal terms representing the assumed trend are introduced. The treatments have to be assigned to plots within blocks in such a way that the known properties of the analysis of variance for the treatment and block sum of squares are preserved and variation due to the trend may be removed from the error sum of squares. When such a design exists, it is called trend-free block design (TF_pB).

7.8.1 The model

Let the plot positions in a block be indexed by m-dimensional vectors of positive integers, $\mathbf{t} = (t_1, \ldots, t_m)'$, $t_u = 1, \ldots, s_u$; $u = 1, \ldots, m$, where s_u is the number of plot positions in the u-th dimension, $\Pi\, s_u = k$. The polynomial trend extending over plots in each block is a function of plot position $\mathbf{t}$ and is expressed as a linear function of m-dimensional orthogonal polynomials of the form

$$\phi_{\boldsymbol{\alpha}}(\mathbf{t}) = \phi_{(\alpha_1, \ldots, \alpha_m)}(t_1, \ldots, t_m) = \Pi\, \phi_{\alpha_u}(t_u),$$

where $\phi_{\alpha_u}(t_u)$ is one-dimensional orthogonal polynomial of degree α_u, satisfying the orthogonality conditions,

$$\sum_{t_u=1}^{s_u} \phi_{\alpha_u}(t_u) = 0 \tag{7.18}$$

and

$$\sum_{t_u=1}^{s_u} \phi_{\alpha_u}(t_u)\phi_{\alpha_u'}(t_u) = 1, \quad \text{if} \quad \alpha_u = \alpha_u' \tag{7.19}$$

$$= 0, \quad \text{if} \quad \alpha_u \neq \alpha_u'$$

and $u = 1, \ldots, m$. $\alpha_u = \alpha_u'$ implies that $\phi_{\alpha_u} = \phi_{\alpha_u'}$.

Thus the mathematical model is written as

$$y_{j\mathbf{t}} = \mu + \sum_{i=1}^{v} \delta^{i}_{j\mathbf{t}}\, \tau_i + \beta_j + \sum_{\boldsymbol{\alpha} \in \boldsymbol{a}} b_{\boldsymbol{\alpha}}\phi_{\boldsymbol{\alpha}}(\mathbf{t}) + e_{jt}. \tag{7.20}$$

$$j = 1, \ldots, b,$$

where $y_{j\mathbf{t}}$ is the observation on plot position $\mathbf{t}$ of block j, $\boldsymbol{a}$ is an index set of p, m-dimensional, non zero vectors of the form $\boldsymbol{\alpha}$, and $b_{\boldsymbol{\alpha}}$ is the regression co-efficient of $\phi_{\boldsymbol{\alpha}}(\mathbf{t})$. The trend effect on plot position $\mathbf{t}$ is

$$\sum_{\boldsymbol{\alpha} \in \boldsymbol{a}} b_{\boldsymbol{\alpha}}\phi_{\boldsymbol{\alpha}}(\mathbf{t}), \tag{7.21}$$

not depending on the particular plot j. $\delta^{i}_{j\mathbf{t}} = 1$ or 0 according as the treatment i is on plot $(j, \mathbf{t})$ or not, and $e_{j\mathbf{t}}$ are independently and identically distributed $N(0, \sigma^2)$ random variables.

The choice of δ^i_{jt} specifies the particular design. Thus we need to determine these values to construct TFB designs to meet certain criteria. The model (7.20) in matrix notation is

$$\mathbf{y} = \mathbf{X\theta} + \mathbf{e} = \mu \mathbf{1}_{bk} + \mathbf{\Delta}'\boldsymbol{\tau} + \mathbf{D}'\boldsymbol{\beta} + \mathbf{X_b}\,\mathbf{b} + \mathbf{e}, \tag{7.22}$$

where the elements in the vectors and matrices are arranged, for convenience, in lexicographic order. Here **y** and **e** are bk-component vectors with elements y_{jt} and e_{jt} in subscript lexicographic order on $(j, \mathbf{t})$, $\boldsymbol{\tau} = (\tau_1, \ldots, \tau_v)'$,

$$\boldsymbol{\beta} = (\beta_1, \ldots, \beta_b)', \ \mathbf{b} = (b_{\alpha_1}, \ldots, b_{\alpha_p})'$$

$$\boldsymbol{\theta} = (\mu, \boldsymbol{\tau}', \boldsymbol{\beta}', \mathbf{b}')', \ \mathbf{X} = [\mathbf{1} \ \ '\mathbf{D}'\mathbf{X}_b],$$

$$\mathbf{\Delta} = (\mathbf{\Delta}_1' \ldots \mathbf{\Delta}_b'), \ \mathbf{D}' = \mathbf{I}_b \otimes \mathbf{1}_k,$$

$$\mathbf{X}_b = \mathbf{1}_b \otimes \boldsymbol{\phi}$$

where $\mathbf{\Delta}_j$ is the $k \times v$ matrix with δ^i_{jt} as the i-th element in row **t** and $\boldsymbol{\phi}$ is the $k \times p$ matrix with $\phi_{\boldsymbol{\alpha}}(\mathbf{t})$ in row **t** and column $\boldsymbol{\alpha}$, $\boldsymbol{\alpha} \in a$, and the rows of $\mathbf{\Delta}_j$ and $\boldsymbol{\phi}$ being in subscript lexicographic order on **t**. Conditions (7.18), (7.19), (7.21) and the assumption that the design is binary imply

$$\mathbf{1}_k'\boldsymbol{\phi} = \mathbf{0} \tag{7.23}$$

$$\boldsymbol{\phi}'\boldsymbol{\phi} = \mathbf{I}_p \tag{7.24}$$

$$\mathbf{\Delta}_j\mathbf{\Delta}_j' = \mathbf{I}_k, \ j = 1, \ldots, b \tag{7.25}$$

Remark 7.3: It is necessary that the order of orthogonal polynomials defining the trend in (7.20) is less than k. From (7.24) since $\boldsymbol{\phi}$ is $k \times p$, the rank of $\boldsymbol{\phi}$ is p and it is necessary that $p \leqslant k$. For if $k = p$, $\boldsymbol{\phi}$ would be nonsingular contradicted by (7.23). Hence $p < k$.

Bradley and Yeh (1980) obtained a necessary and sufficient condition for a design to be trend free. Using the general analysis of covariance approach of Searle (1975), a design described by (7.22) is trend free relative to the trend effects in that model when the indicated treatment (adjusted, if necessary) and block (unadjusted) sums of squares may be computed orthogonal to trend effects or as though the trend effects were omitted from the model.

Definition 7.5: A block design modelled by (7.22) is trend free relative to the trend in the model if $R(\boldsymbol{\tau}/\mu, \boldsymbol{\beta}, \mathbf{b}) = R(\boldsymbol{\tau}/\mu, \boldsymbol{\beta})$, where $R(\boldsymbol{\tau}/\mu, \boldsymbol{\beta}, \mathbf{b})$ denotes the treatment sum of squares for the model (7.22) and $R(\boldsymbol{\tau}/\mu, \boldsymbol{\beta})$ is the treatment sum of squares for the model like (7.22) with trend effects deleted.

A necessary and sufficient condition for a block design to be trend-free is that

$$\mathbf{\Delta}\mathbf{X}_b = \mathbf{0}. \tag{7.26}$$

For details, reference may be made to Bradley and Yeh (1980).

The design construction problem may now be reformulated. Condition (7.26) is equivalently

$$\sum_j \Delta_j \phi = \Delta_+ \phi = \mathbf{0}, \tag{7.27}$$

where $\Delta_+ = \sum_j \Delta_j$. The matrix Δ_+ has non-negative integer elements with row sums as b and column sums as r_i, $i = 1,\ldots, v$. The TFB-design is constructed if Δ_+ satisfying (7.27) is found first and then if b permutation matrices Δ_j are found such that $\sum_j \Delta_j = \Delta_+$. Condition (7.27) implies

$$\sum_{=1}^{b} \sum_{t_1=1}^{s_1} \ldots \sum_{t_m=1}^{s_m} \delta^i_{jt} \ \phi_{\alpha(t)} = 0,$$

for $i = 1,\ldots, v$ and all $\alpha \in \mathcal{a}$. In any TFB design, the total effect of each trend component over the plots assigned to any treatment is zero.

Considering the model (7.22), the normal equations $\mathbf{X'X\theta} = \mathbf{X'y}$ are

$$\begin{bmatrix} bk & \mathbf{r}' & k\mathbf{1}_b' & \mathbf{0}_{1\times p} \\ \mathbf{r} & \mathbf{R} & \mathbf{N} & \Delta\mathbf{X_b} \\ k\mathbf{1}_b & \mathbf{N}' & k\mathbf{I}_b & \mathbf{0}_{b\times p} \\ \mathbf{0}_{p\times 1} & \mathbf{X}_b'\Delta' & \mathbf{0}_{p\times b} & b\mathbf{I}_p \end{bmatrix} \begin{bmatrix} \mu \\ \boldsymbol{\tau} \\ \boldsymbol{\beta} \\ \mathbf{b} \end{bmatrix} = \begin{bmatrix} G \\ \mathbf{T} \\ \mathbf{B} \\ \mathbf{W} \end{bmatrix} \tag{7.28}$$

where G, $\mathbf{T}$ and $\mathbf{B}$, are as defined in Chapter 2 and $\mathbf{W} = \mathbf{X_b\,y}$ is the vector of trend totals.

The reduced normal equations for treatment effects are

$$\mathbf{A\tau} = \mathbf{Q},$$

where

$$\mathbf{A} = \mathbf{R} - \frac{1}{k}\mathbf{NN}' - \frac{1}{b}\Delta\mathbf{X_b}\mathbf{X_b}'\Delta' \tag{7.29}$$

and

$$\mathbf{Q} = \mathbf{T} - \frac{1}{k}\mathbf{NB} - \frac{1}{b}\Delta\mathbf{X}_b\mathbf{W}.$$

For a TFB design $\Delta\,\mathbf{X_b} = \mathbf{0}$ and $\mathbf{A}$ reduces to

$\mathbf{A}_0 = \mathbf{R} - \frac{1}{k}\mathbf{NN}'$, the matrix of the reduced normal equations for model (2.5) with trend terms omitted. Thus a TFB design is connected if the corresponding block design is connected. We then have the following theorem:

Theorem 7.11: Given model (7.22), a block design and any estimable function $\mathbf{p'\tau}$, and a corresponding TFB design,

(i) $\mathbf{p}'\boldsymbol{\tau}$ is also estimable for the TFB design, and

(ii) $v(\mathbf{p}'\hat{\boldsymbol{\tau}}_0) \leqslant v(\mathbf{p}'\hat{\boldsymbol{\tau}})$,

where $\hat{\boldsymbol{\tau}}_0$ is a solution of $\mathbf{A}\boldsymbol{\tau} = \mathbf{Q}$.

For a connected TFB design, the analysis of variance under model (7.22) is

ANOVA

Source	D.F.	S.S.
Blocks (unadjusted)	$b - 1$	$(\mathbf{B}'\mathbf{B} - G^2/b)/k$
Treatments (adjusted)	$v - 1$	$\mathbf{Q}'\mathbf{A}_0^-\mathbf{Q}$
Trend term 1	1	W_1^2/b
Trend term 2	1	W_2^2/b
⋮		
Trend term p	1	W_p^2/b
Error	$bk - b - p - v + 1$	by subtraction
Total	$bk - 1$	$\mathbf{y}'\mathbf{y} - G^2/bk$

Yeh and Bradley (1983) showed that the necessary and sufficient condition (7.27) for a block design to be trend-free under model (7.22) can be simplified. The existence of permutation matrices $\boldsymbol{\Delta}_1, \ldots, \boldsymbol{\Delta}_b$ satisfying (7.27) is guaranteed by the existence of a suitable matrix $\boldsymbol{\Delta}_+$. Under model (7.27), a TFB design exists if and only if there exists a $k \times v$ matrix $\boldsymbol{\Delta}_+$ of non-negative integers such that

(i) $\boldsymbol{\Delta}_+\mathbf{1} = b\mathbf{1}$,
(ii) $\boldsymbol{\Delta}_+'\mathbf{1} = \mathbf{r} = (r_1, \ldots, r_v)'$, and
(iii) $\boldsymbol{\Delta}_+\boldsymbol{\phi} = \mathbf{0}$.

From condition (7.27) the following two results are immediate:

Theorem 7.12: The existence of a TFB design for a specified set of orthogonal trend components, implies the existence of a TFB design for any subset of those components. Similarly the non-existence of a TFB design for a specified set of orthogonal trend components implies the non-existence for any larger set of components containing the specified set.

Theorem: 7.13: The existence of $TF_{\phi} B(\mathbf{v}, b, k, \mathbf{r})$ and $TF_{\phi} B(\mathbf{v}, b^*, k, \mathbf{r}^*)$ implies the existence of a $TF_{\phi} B(\mathbf{v}, b + b^*, k, \mathbf{r} + \mathbf{r}^*)$.

Yeh and Bradley (1983) proved the following result on the maximum number of orthogonal trend components:

Theorem 7.14: For $p = k - 1$, the existence of a TF_{ϕ} B $(v, b, k, \mathbf{r})$ implies that

i) each r_i is a multiple of k.
ii) treatment i appears in r_i/k blocks in each plot position.

Yeh and Bradley also examined the existence of trend-free block designs for specified trends in one or more dimensions. A necessary and sufficient condition for equi-replicated TFB designs was also obtained. The condition is given in the following theorem:

Theorem 7.15: If $\phi = (\phi_1, \ldots, \phi_p)$ and $\mathbf{r} = r\mathbf{1}$ under model (7.22), then a $TF_{\phi} B(v, b, k, r)$ design exists if and only if there exists a $v \times b$ matrix $\mathbf{Z}$ with non-negative elements such that

i) each column of $\mathbf{Z}$ has the integers 1, ..., k as elements along with $v - k$ zero elements
ii) for any $\alpha = 1, \ldots, p$, the sum of the α-th powers of elements for any single row of $\mathbf{Z}$ is $S_\alpha(k, r)$.

Moreover, the matrix $\mathbf{Z}$ completely specifies a TFB design. Some special results have also been obtained for complete and incomplete block designs.

Sometimes a TFB design does not exist and this provides the motivation for consideration of nearly trend free block (NTFB) designs. Yeh, Bradley and Notz (1985) introduced two criteria for NTFB designs. These have been utilized to develop optimal or highly efficient designs when a TFB design does not exist.

Definition 7.6: A block design under model (7.20) is said to be an NTFB design of Type A if

$$\sum_{\alpha=1}^{p} \sum_{l=1}^{v} \{\sum_{j=1}^{b} \sum_{t=1}^{k} \delta_{jt}^{l} \; \phi_\alpha(t)\}^2$$

is minimum among the class of connected designs with the same incidence matrix.

Definition 7.7: A block design under model (7.20) is said to be an NTFB design of Type B, if among the class of connected designs with the same incidence matrix,

i) $\sum_{\alpha=1}^{p-1} \sum_{i=1}^{v} \{\sum_{j=1}^{b} \sum_{t=1}^{k} \delta_{jt}^{i} \phi_{\alpha}(t)\}^2$ is minimum, and

ii) $\sum_{i=1}^{v} \{\sum_{j=1}^{b} \sum_{t=1}^{k} \delta_{jt}^{i} \phi_{p}(t)\}^2$ is minimum among the designs satisfying (*i*).

The condition (7.27) is a special case of the above NTF criteria. The criterion for an NTFB of Type A requires that the overall treatment arrangement is as orthogonal to all specified trend components as possible. The type B criterion first focuses on the lower degree components of the trend. It is useful in situations when there exist designs completely free of trend effects upto degree $p-1$. It is also useful when the trend anticipated in the model is of degree $p-1$, but a design that is NTF with respect to a trend of degree p is chosen in the experiment to guard against incorrect specification of the trend. Further, if $p=1$, the two definitions are equivalent.

The following Lemmas show a basic property of NTFB designs of Type A and Type B:

Lemma 7.4: For a given block design **N** and a prespecified degree p for the polynomial trend under model (7.20), a design is an NTFB design of type A among the class of connected designs if and only if it maximizes trace **A** in (7.29).

Lemma 7.5: If the treatments are equireplicated under model (7.20) and a TFB design exists for a trend of degree $p-1$, then an NTFB design of Type B exists if and only if there is a $v \times b$ matrix $\mathbf{Z} = (z_{ij})$ such that

i) conditions of Theorem 7.15 are satisfied.

ii) $\sum_{i=1}^{v} \{\sum_{j=1}^{b} z_{ij}^{p} - S_p(k, r)\}^2$ is minimum under (*i*) of this Lemma.

Yeh, Bradley and Notz (1985) also considered construction of NTFB designs for only complete block designs with common within block linear or quadratic trend components, $p = 1$ or 2.

7.9 INCOMPLETE BLOCK DESIGNS FOR BIOLOGICAL ASSAYS

The theory of block designs, developed largely for agricultural experiments, finds applications in many other branches of science and technology. Biological assays (Bio-assays) is one such area where these designs have been used profitably. We discuss here some of the useful work done in this area.

7.9.1 Some preliminaries on bio-assays

A bio-assay is an experiment conducted mainly with the objective to compare the relative efficiency of two or more materials on the basis of the response they produce on living organisms. Here the comparison is made on the basis of two sets of doses, one from the standard preparation and the other from test preparation such that they produce the same response. There can be more than one test preparations for comparison against standard preparations. If s and t denote the doses of standard and test preparations respectively such that they produce the same response, the ratio s/t is a measure of relative potency. A specific purpose of all bio-assays is to estimate the strength of a test preparation (or preparations) relative to the standard preparation in terms of potency alongwith a measure of precision.

If the two preparations involved in an assay contain the same effective ingredient, i.e. the one responsible for producing the response, and all other substances which may be present in the preparations behave as completely inert and thus do not influence the response in any way, then such assays are called Analytical Dilution (AD) assays. We are concerned here with such assays only.

Depending on whether or not it is possible to take direct measures of the amount of doses required to produce some specified response, AD assays are further classified into two broad categories viz. (i) Direct assays, (ii) Indirect assays.

We consider here only the indirect assays based on quantitative responses. The principle involved in these assays is to estimate the dose corresponding to a given response by means of dose-response relationship. This relationship is generally made linear through suitable transformations, commonly known as linearizing or metametric transformations, although non-linear assays like parabolic assays have also been reported in the literature. We, however, restrict here to the linear case. The transformed dose and response are respectively called dose and response metameters. Analysis of variance technique is used to test the linearity. The usual transformations are (i) $\log Z = x$, y and (ii) $Z^{\lambda} = x$, y, where x is the dose metameter, y is the response, Z is the dose in original scale and λ is a constant. If the linearizing transformation is of the first type, the assay is known as Parallel Line (PL) assay, while for the second transformation the assay is called Slope Ratio (SR) assay. The main difference between PL and SR assays is that there is an extra dose, called the blank or zero dose, included in the case of SR assays in addition to the two sets of doses of the two preparations.

Many times several test preparations are to be assayed against one standard preparation. In such cases each preparation could be tested against a standard by conducting a separate small assay for each test preparation or alternatively all test preparations can be included in the same assay and tested against one standard preparation. Such assays are called multiple assays. Multiple assays are not only economical but also provide a more precise estimate of regression co-efficient as it is estimated from combined evidence of all subjects. A comprehensive account of these and other bio-assay problems is available in Finney (1978).

7.9.2 Parallel line assays

In PL assays, $\log Z = x$, y is the linearizing transformation used. Let $y = a_s + bx_s$ be the linear regression equation for standard preparation. Under the fundamental assumption of AD assays, $\rho = Z_s/Z_t$ and taking logarithm we have $x_s = \log \rho + x_t$. Hence the response equation for test preparation is $y = a_t + bx_t$, where $a_t = a_s + b \log \rho$. These two equations show that the two lines are parallel and $\log \rho = (a_t - a_s)/b$. This quantity is estimated by $\log \hat{\rho} = \bar{x}_s - \bar{x}_t - (\bar{y}_s - \bar{y}_t)/b$, where $\bar{x}_s, \bar{x}_t$ denote the mean dose values, $\bar{y}_s$ and $\bar{y}_t$ denote the mean responses of the two preparations and b is the common regression co-efficient. Thus when the response is linearly related to log dose, log potency is estimated as the horizontal distance between the two regression lines. Before estimating the relative potency, it is desirable to test with the help of the data collected from the assay if the two lines are parallel. Such a test is called the validity test. It is also desirable to test if the relationship between response and dose metameter is linear. This gives another validity test.

If the number of doses of two preparations are equal, the PL assay is called symmetrical parallel line (SPL) assay, otherwise the assay is called an asymmetrical parallel line (APL) assay. Finney (1978) has shown that the precision of potency estimates is high and also the labour of arithmetical calculations is less tedious in symmetrical assays. But due to shortage of one material or the other the asymmetrical assays cannot be avoided in practice.

In view of the discussions above, the contrasts of major importance in PL assays are (i) the preparation contrast, L_p, (ii) the combined regression contrast, L_1, and (iii) the parallelism contrast, L_1^*. The first two of these provide an estimate of relative potency while the third is important for making validity test. It is therefore desirable that designs for PL assays permit the estimation of these contrasts with maximum precision.

In PL assay the doses of two preparations (standard and test) are equispaced logarithmically and the response is assumed to depend linearly on log dose. For a PL assay with $n_s + n_t$ doses let $s_1,\ldots,s_{n_s}$ and $t_1,\ldots,t_{n_t}$ denote the doses of standard and test preparations respectively on the log scale and the constant difference between the successive log doses being the same for both the preparations. For notational convenience it is assumed that $s_i < s_{i+1}$ and $t_j < t_{j+1}$, $i = 1,\ldots,n_s - 1$; $j = 1,\ldots,n_t - 1$. The $v - 1$ degrees of freedom can be split into single degree of freedom contrasts L_p, L_h and L_h^*, $h = 1, 2,\ldots$ Here L_h and L_h^* denote the sums and differences of h-th power regression. It is useful to further subgroup L_h and L_h^* into odd and even numbered contrasts L_{2n+1}, L_{2n+2}, L_{2n+1}^*, L_{2n+2}^*, respectively.

Consider an incomplete block design for a PL assay in $v(= n_s + n_t)$ doses arranged in b blocks such that the j-th block receives k_{sj} doses of standard and k_{tj} doses of test preparations. Clearly $k_j = k_{sj} + k_{tj}$, $j = 1,\ldots,b$. Let each dose of standard preparation appears in r_s blocks and each dose of test preparation appears in r_t blocks. We also assume the usual intra block model with doses of standard and test preparations representing the treatment effects. It is known (see Finney, 1978) that the estimate of contrast L_p is given by

$$\hat{L}_p = \sum_{i=1}^{n_s} S_i/N_s - \sum_{j=1}^{n_t} T_j/N_t, \tag{7.30}$$

where $S_i(T_i)$ is the total of observations from the i-th dose of standard (test) preparation, $N_s = r_s n_s$, $N_t = r_t n_t$. Again, the combined regression contrast L_1 is estimated by

$$c\hat{L}_1 = \sum_{i=1}^{n_s}\left(i - \frac{n_s+1}{2}\right)S_i + \sum_{j=1}^{n_t}\left(j - \frac{n_t+1}{2}\right)T_j \tag{7.31}$$

and the parallelism contrast L_1^* is estimated by

$$\hat{L}_1^* = \sum_{i=1}^{n_s}\left(i - \frac{n_s+1}{2}\right)S_i/S_{xx} - \sum_{j=1}^{n_t}\left(j - \frac{n_t+1}{2}\right)T_j/T_{xx}, \tag{7.32}$$

where

$$c = S_{XX} + T_{XX}$$

and

$$T_{XX} = r_t n_t\,(n_t^2 - 1)/12,$$
$$S_{XX} = r_s n_s\,(n_s^2 - 1)/12.$$

Let $\mathbf{e}_p$, $\mathbf{e}_{2n+1}$, $\mathbf{e}_{2n+1}^*$, $\mathbf{e}_{2n+2}$, and $\mathbf{e}_{2n+2}^*$ denote respectively the vectors of co-efficients of the bio-assay contrasts L_p, L_{2n+1}, L_{2n+1}^*, L_{2n+2}, and L_{2n+2}^*, respectively. Assuming $r_s = r_t = r$ and further that the doses are arranged in the order $a_1, \ldots, a_{mj}, a_{2mj}, a_{2mj-1}, \ldots, a_{mj+1}$ if $n_s = 2m_j$ and $a_1, \ldots, a_{mj}, a_{2mj+1}, a_{2mj}, \ldots, a_{mj+1}$, if $n_j = 2m_j + 1$, where $a_l = s_l$ or t_l and $j = s$ or t, the co-efficient vectors can be expressed as

$$\begin{aligned}
\mathbf{e}_p &= [n_t\mathbf{1}'_{n_s}, -n_s\mathbf{1}'_{n_t}]',\\
\mathbf{e}_{2n+1} &= [\mathbf{x}'_{n_s}, \mathbf{x}'_{n_t}]',\\
\mathbf{e}_{2n+1}^* &= [\mathbf{x}'_{n_s}, -\mathbf{x}'_{n_t}]'\\
\mathbf{e}_{2n+2} &= [\mathbf{y}'_{n_s}, \mathbf{y}'_{n_t}]'\\
\mathbf{e}_{2n+2}^* &= [\mathbf{y}'_{n_s}, -\mathbf{y}'_{n_t}]'
\end{aligned} \tag{7.33}$$

where $\mathbf{x}_{nj}$ and $\mathbf{y}_{nj}$ are n_j-component column vectors satisfying $\mathbf{x}'_{nj}\mathbf{1} = \mathbf{y}'_{nj}\mathbf{1} = 0$, $j = s$ or t. Here $\mathbf{x}_{nj}$ and $\mathbf{y}_{nj}$ are orthogonal polynomials of degree $2n+1$ and $2n+2$ respectively. The co-efficient vectors in (7.33) can also be expressed as

$$\mathbf{1}_{nj} = (1, 1)' \otimes \mathbf{1}_{mj},\ \mathbf{x}_{nj} = (1, -1)' \otimes \mathbf{x}_{mj},$$
$$\mathbf{y}_{nj} = (1, 1)' \otimes \mathbf{y}_{mj} \text{ if } n_j = 2m_j,$$

and

$$\begin{aligned}\mathbf{1}_{n_j} &= [(1, 1)' \otimes \mathbf{1}'_{m_j}, 1]',\\ \mathbf{x}_{n_j} &= [(1, 1)' \otimes \mathbf{x}'_{m_j}, 0]', \qquad (7.34)\\ \mathbf{y}_{n_j} &= [(1, 1)' \otimes \mathbf{y}'_{m_j}, Z_j]', \text{ if } n_j = 2m_j + 1,\end{aligned}$$

where $\mathbf{x}_{m_j}$ and $\mathbf{y}_{m_j}$ are m_j-component vectors with $\mathbf{y}'_{m_j}\,\mathbf{1} = 0$ if n_j is even and $\mathbf{y}'_{m_j}\,\mathbf{1} \neq 0$ if n_j is odd and Z_j is a scalar constant.

7.9.3 Slope ratio assays

In slope ratio assays response is a linear function of some power of dose say $Z^\lambda = x, y$. Let there be n_s and n_t doses of standard and test preparations, respectively and let there be a blank dose. If $n_s = n_t$, the assay is called a symmetrical slope ratio (SSR) assay, otherwise the assay is asymmetrical slope ratio (ASR) assay. In SR assays, the doses are taken in arithmetical progression, i.e. $s_i = i$, $i = 1,\ldots,n_s$ and $t_j = j$, $j = 1,\ldots,n_t$. For a $(n_s + n_t + 1)$ point assay, between doses sum of squares with $n_s + n_t$ degrees of freedom can be split into sum of squares due to regression (β_s and β_t) with one degree of freedom each, blank (L_B), intersection (L_I), sums and differences of h-th power contrasts for $h = 2, 3,\ldots$, each with one degree of freedom. L_B and L_I are useful for making validity tests and β_s and β_t are useful for estimation of relative potency. The component L_1 should be significant while L_B, L_I and the remaining components should be non-significant in order that the assay is valid. If $y = a + b_s x_s$ is the relationship for standard preparations and because $Z_s = \rho Z_t$, the relationship for the test preparation on the assumption of similarity of response is $y = a + b_t x_t$, where $b_t = \rho^\lambda b_s$. Similarity in response requires that the two regression lines intersect on response axis. For testing this a blank dose is included. As these two regression lines meet at response axis, the relative potency ρ is estimated by $\hat{\rho} = (b_t/b_s)^{1/\lambda}$. As a measure of precision of the estimate of relative potency, fiducial limits at required probability level are obtained using Fieller's theorem. A comprehensive account of this and other related problems is available in Finney (1978).

Consider an incomplete block design for SR assay in $v = n_s + n_t + 1$ points and b blocks. Each dose of standard and test preparations is replicated r times whereas the control (blank dose) is included in each of the b blocks making its replication b. Assuming the usual intra block model and $n_s = n_t = n$, Finney (1978) has shown that the estimates of L_B and L_I are

$$\hat{L}_B = gc + \sum_{i=1}^{n} [3(i-1) - h]\,(S_i + T_i), \qquad (7.35)$$

$$\hat{L}_I = \sum_{i=1}^{n} [h - 3(i-1)]\,(S_i - T_i), \qquad (7.36)$$

where $S_i(T_i)$ is the total of observations from the i-th dose of standard (test) preparations, c is the total of observations from blank dose, $g = n(n-1)$ and $h = 2(n-1)$.

7.9.4 Characterization of designs for bio-assays

It is seen earlier that in bio-assays not all contrasts are of equal importance and it is desired to estimate contrasts used in validity tests with maximum precision. We now study some patterns of designs which enable these contrasts to be estimated with full precision The results reported here are from Kyi Win and Dey (1980), Gupta, Nigam and Puri (1987) and Puri, Gupta and Nigam (1987).

7.9.5 Desirable patterns of designs for parallel line assays

Consider an incomplete block design for a PL assay and assume the usual intra-block model. For a complete block design,

$$E(\hat{L}_p) = \bar{s} - \bar{t} = L_p,$$

so that L_p can be estimated through $\hat{L}_p$ free from block effects, where

$$\bar{s} = \sum_{i=1}^{n_s} s_i/n_s,\ \bar{t} = \sum_{j=1}^{n_t} t_j/n_t,$$

and $\hat{L}_p$ is as given in (7.30). However, if we consider an incomplete block design, then we have

$$E(\hat{L}_p) = \bar{s} - \bar{t} + \sum_{j=1}^{b} (k_{sj}/N_s - k_{tj}/N_t)b_j. \tag{7.37}$$

From (7.37) it is easily seen that $\hat{L}_p$ estimates L_p free from block effects if

$$k_{sj}/N_s = k_{tj}/N_t,\ j = 1,\ldots, b. \tag{7.38}$$

If, however, $r_s = r_t = r$, (7.38) boils down to

$$k_{sj}/k_{tj} = n_s/n_t,\ j = 1,\ldots, b.$$

Again, it can be shown that $\hat{L}_1$, given in (7.31), estimates L_1 free from block effects if

$$\sum_j{}' \left(i - \frac{n_s + 1}{2}\right) = 0 = \sum_j{}'' \left(i - \frac{n_t + 1}{2}\right). \tag{7.39}$$

Further, under (7.39), $\hat{L}_1^*$ given in (7.32), estimates the parallelism contrast L_1^* free from block effects. Thus L_1 and L_1^* can be estimated orthogonally if

i) $\sum_j{}' i = k_{sj}(n_s + 1)/2 \quad (7.40)$

ii) $\sum_j'' i = k_{t_j}(n_t + 1)/2,\ j = 1, \ldots, b,$

where $\sum_j' i$ and $\sum_j'' i$ are summation for standard and test preparations dose indices respectively in the j-th block.

Conditions (7.38) and (7.40) were obtained by Kyi Win and Dey (1980). We now describe another characterization of such designs.

Consider again an incomplete block design for PL assay with $r_s = r_t = r$. The $v \times b$ incidence matrix is

$$\mathbf{N} = \begin{bmatrix} \mathbf{N}_s \\ \mathbf{N}_t \end{bmatrix} \tag{7.41}$$

where $\mathbf{N}_j$ ($j = s$ or t) is an $n_j \times b$ incidence matrix of standard (test) preparation. The matrix $\mathbf{M} = \mathbf{N}\mathbf{K}^{-1}\mathbf{N}'/r$ is of the form

$$\mathbf{M} = \begin{bmatrix} \mathbf{D}_{ss} & \mathbf{D}_{st} \\ \mathbf{D}_{ts} & \mathbf{D}_{tt} \end{bmatrix} \tag{7.42}$$

where

$$\mathbf{D}_{jj} = \mathbf{N}_j\mathbf{K}^{-1}\mathbf{N}_j'/r,\ j = s \text{ or } t$$

and

$$\mathbf{D}_{jj'} = \mathbf{N}_j\mathbf{K}^{-1}\mathbf{N}_{j'}'/r,\ j \neq j' = s \text{ or } t.$$

The contrasts L_p, L_{2n+1} and L_{2n+1}^* will be estimated free from block effects if $\mathbf{e}_p$, $\mathbf{e}_{2n+1}$ and $\mathbf{e}_{2n+1}^*$ are eigen vectors of matrix $\mathbf{M}$ corresponding to eigen value zero. We then have the following theorem:

Theorem 7.16: For orthogonal estimation of L_p, L_{2n+1} and L_{2n+1}^*, the matrix $\mathbf{M}$ in (7.42) should be such that

i) $\mathbf{D}_{jj'}\mathbf{1}_{n_{j'}} = \{n_{j'}/(n_s + n_t)\}\ \mathbf{1}_{n_{j'}}$

ii) $\mathbf{D}_{jj'}\mathbf{x}_{n_{j'}} = \mathbf{0},\ j, j' = s$ or t. (7.43)

Proof: $\mathbf{M}\mathbf{e}_p = \mathbf{0}$ and $\mathbf{M}\mathbf{1} = \mathbf{1}$ gives (i). Similarly $\mathbf{M}\mathbf{e}_{2n+1} = \mathbf{0}$ and $\mathbf{M}\mathbf{e}_{2n+1}^* = \mathbf{0}$ gives (ii).

We now obtain the structure of $\mathbf{D}_{jj'}$ matrices in (7.42) such that (7.43) is satisfied. In view of the nature of coefficient vectors of the contrasts, it is not unrealistic to assume

$$\mathbf{D}_{jj'} = \mathbf{J}_2 \otimes \mathbf{B}_{jj'},\ j, j' = s \text{ or } t, \tag{7.44}$$

when

$$n_j = 2m_j$$

and

$$\mathbf{D}_{jj} = \begin{bmatrix} \mathbf{J}_2 \otimes \mathbf{B}_{jj} & \mathbf{1}_2 \otimes \boldsymbol{\phi}_j \\ \mathbf{1}_2' \otimes \boldsymbol{\phi}_j' & \alpha_j \end{bmatrix} \tag{7.45}$$

$$\mathbf{D}_{jj'} = \begin{bmatrix} \mathbf{J}_2 \otimes \mathbf{B}_{jj'} & \mathbf{1}_2 \otimes \boldsymbol{\xi}_j \\ \mathbf{1}_2' \otimes \boldsymbol{\xi}_{j'}' & \beta_{j'} \end{bmatrix} \tag{7.46}$$

when

$$n_j = 2m_j + 1, j \neq j' = s \text{ or } t.$$

The matrix $\mathbf{B}_{jj'}$ is of order $m_j \times m_{j'}$, $\boldsymbol{\phi}_j$ and $\boldsymbol{\xi}_j$ are m_j-component column vectors and α_j and β_j are scalar constants. Notice that

$$\mathbf{B}_{jj'}\mathbf{1}_{m_{j'}} = \{n_{j'}/2(n_s + n_t)\}\mathbf{1} \text{ if } n_j = 2m_j$$

and

$$\mathbf{B}_{jj}\mathbf{1}_{mj} + (1/2)\boldsymbol{\phi}_j = [n_j/2(n_s + n_t)]\mathbf{1}$$

$$\boldsymbol{\phi}_j'\mathbf{1}_{mj} + (1/2)\alpha_j = [n_j/2(n_s + n_t)]$$

if $n_j = 2m_j + 1$. Similar relations hold for $\mathbf{B}_{jj'}$, $\boldsymbol{\xi}_{j'}$, and $\beta_{j'}$.

Remark 7.3: In SPL assays it may be convenient to work with $\mathbf{D}_{11} = \mathbf{D}_{22} = \mathbf{D}$, because it simplifies the method of construction of designs. The structure of matrices $\mathbf{D}$ and $\mathbf{D}_{12}$ satisfying (7.43) simplifies to

$$\mathbf{D} = \mathbf{J}_2 \otimes \mathbf{B}_1,\ \mathbf{D}_{12} = \mathbf{J}_2 \otimes \mathbf{B}_2 \tag{7.47}$$

if $n = 2m$ and

$$\mathbf{D} = \begin{bmatrix} \mathbf{J}_2 \otimes \mathbf{B}_1 & \boldsymbol{\delta} \\ \boldsymbol{\delta}' & \alpha \end{bmatrix} \tag{7.48}$$

$$\mathbf{D}_{12} = \begin{bmatrix} \mathbf{J}_2 \otimes \mathbf{B}_2 & \boldsymbol{\delta} \\ \boldsymbol{\delta}' & \alpha \end{bmatrix} \tag{7.49}$$

when $n = 2m + 1$. Here the matrix $\mathbf{B}_l$ is of order $m \times m$ such that $\mathbf{B}_1\mathbf{1} = \mathbf{B}_2\mathbf{1}$, $\boldsymbol{\delta}$ is a $2m$-component column vector of non-negative real numbers and α is a positive scalar constant and is a function of design parameters.

We now prove an interesting result on designs with matrix $\mathbf{M}$ as in (7.42) which shows that the loss of information on various contrasts can be worked out from the component designs only.

Theorem 7.17: In SPL assays ($n_s = n_t = n$) with matrix $\mathbf{M}$ as in (7.42), the losses of information on L_{2n+2} and L^*_{2n+2} is $2(\alpha + \beta)/r$ and $2(\alpha - \beta)/r$, respectively if

i) $\mathbf{D}\mathbf{x}_n = \alpha\mathbf{x}_n$

(7.50)

ii) $\mathbf{D}_{12}\mathbf{y}_n = \beta\mathbf{y}_n$

Proof: The proof follows from considering $\mathbf{M}\mathbf{e}_{2n+2}$ and $\mathbf{M}\mathbf{e}^*_{2n+2}$.

Corollary 7.12: If $\alpha = \beta$ (or $\alpha = -\beta$), then L^*_{2n+2} (or L_{2n+2}) also becomes free from block effects and loss of information on L_{2n+2} (or L^*_{2n+2}) is $4\alpha/r$.

From the characterizations of Theorem 7.16 and Theorem 7.17 it follows that $\mathbf{e}_p$, $\mathbf{e}_{2n+1}$, $\mathbf{e}^*_{2n+1}$, $\mathbf{e}_{2n+2}$ and $\mathbf{e}^*_{2n+2}$ are basic contrasts and therefore the designs for SPL assays with these characterizations are PEB with three efficiency classes, and losses 0, $2(\alpha + \beta)/r$ and $2(\alpha - \beta)/r$, respectively. When $\alpha = \beta$ (or $\alpha = -\beta$), there are two losses of information 0 and $4\alpha/r$ and the design is SPEB.

7.9.6 Designs for biological assays

It has been stated earlier that in bio-assays some contrasts need to be estimated with maximum precision. This is not possible by using the conventional block designs like BIB and PBIB. These designs have to be modified before they can be used for bio-assays. Das and Kulkarni (1966) proposed two series of incomplete block designs for SPL assays. These designs are obtained by modifying BIB and circular designs. These designs permit estimation of L_p and L_1 contrasts with full information. These contrasts are useful in the estimation of relative potency. However, the contrast L_1^* required for fundamental validity test cannot be estimated free from block effects through these designs. Kulshrestha (1971) constructed three series of incomplete block designs for SPL assays by suitably modifying a BIB or PBIB or circular design and termed these designs as modified incomplete block bio-assay designs. These designs permit the estimation of all the important contrasts with full efficiency but are available only for even number of doses with minimum block size eight. Following the same approach Seshagiri (1974) gave three series of incomplete block designs for PL assays. The parallelism contrast required for making validity tests cannot be estimated without loss of information through these designs. Kyi Win and Dey (1980) developed designs for PL assays by using conditions (7.38) and (7.40) required for orthogonal estimation of L_p, L_1 and L_1^* contrasts. Designs for 6 to 20 points with a maximum of five replications are also catalogued by them. Recently, Nigam and Boopathy (1985) gave a systematic method of constructing designs for SPL assays. These designs permit orthogonal estimation of L_p, L_{2n+1} and L_{2n+1} contrasts.

For some further interesting methods on the construction of designs for PL assays reference may be made to Gupta, Nigam and Puri (1987), and Puri, Gupta and Nigam (1987). The construction of designs for SR assays does not pose any new problems. These designs can be obtained from designs for PL assays by including a blank dose in each block. The resulting design is a supplemented block design and can be analysed easily. Das and Kulkarni (1966) constructed a series of incomplete block designs for SSR assays obtained by suitably modifying BIB designs. These designs permit orthogonal estimation of intersection contrast L_1. These designs were termed as modified BIB designs. Kulshrestha (1971) proposed two series of incomplete block designs for SSR assays for even number of doses of each preparation.

But these designs provide shorter fiducial intervals for the relative potency than those obtained through the use of Das and Kulkarni designs. Finney (1978) pointed out that the shorter fiducial intervals come from the effectively greater replication. Other relevant references on construction of designs for SR assays are those of Seshagiri (1974), Kyi Win (1978) and Boopathy (1980).

8. Other Considerations in Block Designs

8.1 INTRODUCTION

So far the discussions have been limited to obtaining characterizations of block designs which admit simplified analysis. Once the design has been identified as an EB or PEB or SPEB, the analysis is straightforward and the loss of information on treatment contrasts is known the moment we have the knowledge of the design. However, mere simplicity in statistical computations does not necessarily guarantee that the design is good, although quite often a design developed from intuitive reasoning turns out to be good according to some property or a given measure of goodness. A measure of goodness of a design is optimality which must be given consideration next only to ease in computations.

The subject of optimal designs is discussed in Section 8.2. Some other properties of block designs considered in the literature are those of robustness and resistance. We discuss these in Sections 8.3 and 8.4 respectively. In Section 8.5 Grubb's estimators are described which may be useful when there is some departure from the usual assumptions of homogeneity of error variances.

8.2 OPTIMAL DESIGNS

The work on optimal designs was initiated by Smith (1918) and Wald (1943) and later on developed to a great extent by Kiefer in a series of papers starting from 1958. Recently Cheng has also contributed greatly to the optimal designs of experiments. Many computer aided algorithms have also been developed for obtaining optimal designs.

We restrict to the usual linear model (1.1) and the reduced normal equations (1.2). It is well known that a linear parametric function $\boldsymbol{\xi} = \mathbf{l}'\boldsymbol{\theta}$, is estimable under a design d if $\mathbf{l}'$ is in the row space of matrix $\mathbf{A}$ and the best linear unbiased estimator of $\boldsymbol{\xi}$ is given by $\hat{\boldsymbol{\xi}} = \mathbf{l}'\mathbf{A}^{-}\mathbf{Q}$ with $Var(\hat{\boldsymbol{\xi}}) = \mathbf{l}'\mathbf{A}^{-}\mathbf{l}\sigma^2$. In this case the design d is connected for $\boldsymbol{\xi}$. Now suppose that the design d is connected for a set of $v-1$ independent orthonormal contrasts $\mathbf{P}'\boldsymbol{\theta}$, where $\mathbf{P}$ is a $v \times (v-1)$ matrix whose columns belong to the row space of matrix A. Then the covariance matrix of the best linear unbiased estimate of $\mathbf{P}'\boldsymbol{\theta}_1$ is given by

$$\mathbf{V}_d\sigma^2 = \mathbf{P}'\mathbf{A}^-\mathbf{P}\sigma^2 = (\mathbf{P}'\mathbf{A}\mathbf{P})^-\sigma^2.$$

This leads to the consideration of an optimality functional ψ on $(v-1) \times (v-1)$ matrices and to the determination of a design d which minimizes $\psi(\mathbf{V}_d)$. A design solving this problem is said to be ψ-optimal. ψ is called an optimality criterion. Some commonly used optimality criteria are

D-optimality : $\psi(\mathbf{V}_d) = |\mathbf{V}_d|$, determinant $\mathbf{V}_d$,

A-optimality : $\psi(\mathbf{V}_d) =$ trace $\mathbf{V}_d$,

E-optimality : $\psi(\mathbf{V}_d) =$ maximum eigen value of $\mathbf{V}_d$.

All these optimality criteria have a simple geometric meaning in the case in which one might give an ellipsoidal confidence set instead of point estimate. The determinant is proportional to the volume of the ellipsoid, the trace to the sum of squares of lengths of the principal axes and the maximum eigen value to the square of the maximum diameter.

The relationship between these optimality criteria is well known and may be found in Kiefer (1958, 1959, 1975). In some setting it is possible to introduce an optimality criterion which includes D-, A- and E-optimality as special cases. One such setting together with a sufficient condition under which a design is optimal is given by Kiefer (1975).

A major difficulty is the computation of $(\mathbf{P}'\mathbf{A}\mathbf{P})^-$ for each competing design d. To avoid such computations one may utilize the result of Kiefer (1975) on universal optimality. The concept of universal optimality is the following:

Assume that $\mathbf{R}_v$ consists of $v \times v$ non-negative definite matrices and $\overline{\mathbf{R}}_v$ consist of those elements of $\mathbf{R}_v$ all of whose row and column sums are zero. Let Ω be the set of all functions ω from $\overline{\mathbf{R}}_v$ to $[-\infty, +\infty]$ with the following properties:

i) $\omega(\cdot)$ is convex,

ii) $\omega(b\mathbf{R})$ is non-increasing in the scalars $b \geqslant 0$, $\mathbf{R} \in \overline{\mathbf{R}}_v$ (8.1)

iii) $\omega(\cdot)$ is invariant under each simultaneous permutation of rows and columns of $\mathbf{R} \in \overline{\mathbf{R}}_v$.

Under this setting, we now define a very useful concept of optimality.

Definition 8.1: A design d^* is universally optimal in the class of competing designs under consideration if $\omega(\mathbf{A}_{d^*}) \leqslant \omega(\mathbf{A}_d)$ for each $\omega \in \Omega$.

If d^* is universally optimal, then it is D-, A- and E-optimal. In some situations it is possible to identify the universally optimal design without actually computing $\boldsymbol{\omega}(\mathbf{A}_d)$ for each competing design. One such situation identified by Kiefer (1975) is the following:

Definition 8.2: A design d is said to be completely symmetric with respect to $\boldsymbol{\theta}_1$ if its corresponding matrix $\mathbf{A}_d$ is of the form $a\mathbf{I}+b\mathbf{1}\mathbf{1}'$, for scalar constants a and b.

If the class of competing designs contains a design d^* such that

i) $\mathbf{A}_{d^*}$ is completely symmetric,

ii) trace $(\mathbf{A}_{d^*}) \geqslant$ trace $(\mathbf{A}_d)$ for all $d \in \bar{\mathbf{R}}_v$, (8.2)

then d^* is universally optimal.

Remark 8.1: Since $-$trace **(A)** satisfies (8.1) it follows that (8.2) (ii) is necessary for universal optimality.

Remark 8.2: If a class of matrices $A = \{\mathbf{A}_d, d \in \mathcal{D}\}$ contains an $\mathbf{A}_{d^*}$ which is a multiple of $\mathbf{I}_v$ and maximizes trace $\mathbf{A}_d$ for $d \in \mathcal{D}$, then d^* is universally optimal.

8.2.1 Optimal block designs

Consider a block design $D(v, b, \mathbf{r}, \mathbf{k})$. Kshirsagar (1958), Mote (1958) and Roy (1958) established the optimality of balanced incomplete block designs in different classes of designs. Kiefer (1958) established the *A*-, *D*- and *E*-optimality of balanced block designs (BBD) for infering on any full set of orthonormal treatment contrasts. A BBD is a block design with v treatments, b blocks, block size k and incidence matrix $\mathbf{N} = (n_{ij})$ satisfying the following requirements:

i) $r_i = \sum_{j=1}^{b} n_{ij} = r$ for all $i = 1,\ldots, v$,

ii) $\lambda_{ih} = \sum_{j=1}^{b} n_{ij}n_{hj} = \lambda$ for all $i \neq h = 1,\ldots, v$,

iii) $|n_{ij} - k/v| \leqslant 1$.

Obviously a BIB design is a special case of BBD. A BBD is necessarily variance balanced. Dey and Gupta (1986) have shown that it is not generally true that a variance balanced design is always better than a comparable EB design. If we restrict the class of competing designs to a subclass of $D(v, b, \mathbf{r}, k)$, then it is sometimes possible to construct EB designs which are superior to the corresponding variance balanced designs, according to say *A*-, *D*- or *E*-optimality criterion. The following example illustrates this point. Let D_0 be the class of all designs with $v = 4$, $b = 10$, $k = 3$, and replication vector $\mathbf{r} = (6, 6, 6, 12)'$. Consider the designs d_1 and d_2 belonging to D_0:

$$d_1\text{: } (1, 2, 3);\ (1, 2, 4);\ (1, 2, 4);\ (1, 3, 4);\ (1, 3, 4);$$
$$(1, 4, 4);\ (2, 3, 4);\ (2, 3, 4);\ (2, 4, 4);\ (3, 4, 4)$$
$$d_2\text{: } (1, 2, 3);\ (1, 2, 4);\ (1, 3, 4);\ (2, 3, 4);\ (1, 2, 3);$$
$$(1, 2, 4);\ (1, 3, 4);\ (2, 3, 4);\ (4, 4, 4);\ (4, 4, 4).$$

It can be easily verified that the EB design d_1 is superior to the variance balanced design D_2 on the basis of A- and D-optimality criteria for infering on a complete set of orthonormal contrasts. In situations when a BBD or a BIB design does not exist some results on optimality of block designs have been obtained by Takeuchi (1961), Sinha (1980) and Conniffe and Stone (1974, 75), who showed that a GD design with $\lambda_2 = \lambda_1 + 1$ is E-optimal if there are only two groups. Further results in this direction are due to Cheng (1978, 1979, 1980) and John and Mitchell (1977).

From a purely intuitive appeal, Shah (1960) suggested an optimality criterion applicable to a class of designs for which trace **A** is constant. Suppose that $\lambda_1,\ldots,\lambda_{v-1}$ are the non-zero eigen values of **A** such that trace $\mathbf{A} = \Sigma\lambda_i$ is a constant. Shah's criterion is to minimize the quantity $\Sigma(\lambda_i - \bar{\lambda})^2/(v-1)$, where $\bar{\lambda}$ is the average of non-zero eigen values of **A**. Such designs are called S-optimal designs. Actually Shah's criterion requires minimization of trace $\mathbf{A}^2$. Although the S-optimality criterion does not have any appealing statistical significance, in many situations it gives designs which are good according to other optimality criteria as well. Removing the restriction that trace **A** is constant, Ecclesston and Hedayat (1974) defined another optimality criterion which requires the maximization of trace **A** at the first stage and minimization of trace $\mathbf{A}^2$ at the second stage. Such designs are called (M-S) optimal designs. Usually (M-S) optimal designs are not unique.

The properties of proper (M-S) optimal block designs have been studied by Jacroux (1978). He showed that (i) for given v, b and k, if $bk \geqslant v+b-1$, there always exists a connected (M-S)-optimal design with the given parameters, (ii) the complement of any proper, binary (M-S) optimal block design is (M-S) optimal and (iii) the dual of any proper, equi-replicate, (M-S) optimal design is (M-S) optimal.

A computer algorithm for constructing proper, (M-S) optimal, incomplete block designs has been developed by Nam (1983). This algorithm is then used to construct (M-S) optimal incomplete block designs (v, b, k) with $v = 4$, 5 and 6, $2 \leqslant k \leqslant v-2$, and for different b. The efficiencies of some of these designs are then compared with corresponding nearly balanced incomplete block designs (NBIBD) of John and Mitchell (1977) and Cheng and Wu (1981). These authors constructed NBIBDs through the graph theoretic approach. Cheng (1978) established the (M-S) optimality of a regular graph design. Further, if there is a regular graph design, then any (M-S)-optimal design should also be a regular graph design.

A regular graph design is an incomplete block design satisfying the following conditions:

i) Each treatment appears in each block at most once.

ii) All the treatments are equally replicated.

iii) Let λ_{dij} be the number of times treatments i and j appear together in a design d. Then $|\lambda_{dij} - \lambda_{di'j'}| \leqslant 1$ for all $i \neq j$, $i' \neq j'$.

Yeh, Bradley and Notz (1985) established the A- and D-optimality of NTFB designs. They showed that when the model (7.20) is based on a BIB or a complete block design with incidence matrix **N** under a polynomial trend of degree p, an NTFB design of Type B is A- and D-optimal among the class of trend free designs with the same incidence matrix under a trend with degree $p-1$. When $p = 1$, and for given N, all of the following are equivalent: (i) NTF of Type A, (ii) NTF of Type B, (iii) A-optimal (iv) D-optimal. Bradley and Yeh (1980) have shown that if a TFB design exists in association with a given block design, specified **N**, it is both A- and D-optimal.

8.3 ROBUST DESIGNS

Consider the usual linear model (1.1). Let the column rank of matrix **X** be m ($\leqslant k$), where **X** is an $n \times k$ matrix. Suppose that the u-th observation is an outlier. In other words, in place of y_u, the observation recorded is $y_u + c$. Assuming **X′X** to be of full rank, Box and Draper (1975) showed that the predicted response $\hat{\mathbf{y}}$ at any given point will be insensitive to the outlier if the quantity $r = \sum_u r_{uu}^2$ is minimized, where r_{uu} is the u-th diagonal element of the matrix $\mathbf{R} = \mathbf{X}(\mathbf{X}'\mathbf{X})^{-1}\mathbf{X}'$. Designs minimizing r are called robust designs.

The assumption of full rank of the **X′X** restricts the applicability of the Box-Draper criterion to response surface designs only. In the block designs the matrix **X** does not have full column rank. For such a situation Gopalan and Dey (1976) developed a criterion of robustness on lines similar to those given by Box and Draper.

In the usual Gauss-Markoff set up, $R_0^2/(n-m)$ is an unbiased estimate of σ^2 in the absence of outliers. When the u-th observation y_u has added to it a quantity c, the discrepancy in estimating σ^2 through R_0^{*2} is $d_u = c^2 a_{uu}/(n-m)$, where R_0^{*2} is the residual sum of squares in the presence of the outlier and a_{uu} is the u-th diagonal element of $\mathbf{S}^* = \mathbf{I} - \mathbf{X}(\mathbf{X}'\mathbf{X})^{-1}\mathbf{X}'$. Assuming that the discrepancy c could occur with any one of the observations, the corresponding discrepancies would be $d_1,\ldots, d_n$ with average discrepancy $\bar{d} = c^2/n$. Thus the average discrepancy is fixed for a given number of observations. In order that no unduly large discrepancy in the estimator of σ^2 is caused by the outlier, $d_u(u = 1,\ldots,n)$ should be made as uniform as possible in view of the fact that $\bar{d}$ is fixed. Box and Draper (1975) pointed out that one measure of uniformity is the variance of d, given by $V(d)=c^4 (s - m^2/n)/\{n(n - m)^2\}$, where $s = \sum_u s_{uu}^2$, and s_{uu} is the u-th diagonal element of $\mathbf{S} = \mathbf{X}(\mathbf{X}'\mathbf{X})^{-}\mathbf{X}'$. For fixed n and m, minimization of $V(d)$ implies that s should be minimized which in turn requires that all s_{uu}'s should be equal, since $\sum_u s_{uu} = m$ is fixed. A design giving equal values of s_{uu} may be called robust. When $m = k$, the s-and r-criterion are identical.

From the preceding discussion it is obvious that the study of robust designs requires the computation of the diagonal elements of $\mathbf{S}$. Since the dispersion matrix of the least square estimates $\mathbf{X}\,\hat{\boldsymbol{\theta}}$ of $\mathbf{X}\,\boldsymbol{\theta}$ is $\sigma^2\,\mathbf{S}$, a design will be robust if and only if all the components of $\mathbf{X}\boldsymbol{\theta}$ are estimated with the same variance. For block designs the u-th component of $\mathbf{X}\,\boldsymbol{\theta}$ is $(\mu + \tau_i + \beta_j)$, where the u-th observation pertains to the i-th treatment in j-th block. The least square estimate of $\mu + \tau_i + \beta_j$ is

$$p_{ij} = \hat{\tau}_i + B_j/n_{\cdot j} - \sum_i n_{ij}\,\hat{\tau}_i/n_{\cdot j},$$

where B_j is the j-th block total, $j = 1,\ldots, b$, n_{ij} is the number of times the i-th treatment appears in the j-th block, $i = 1,\ldots, v$, $\sum_i n_{ij} = n_{\cdot j}$, and $\hat{\tau}_i$ is any solution of the reduced normal equations $\mathbf{A}\boldsymbol{\tau} = \mathbf{Q}$. Thus any design will be robust if and only if $V(p_{ij})$ is a constant independently of i and j. This is a useful idea in characterizing robust designs. It is assumed that the design is connected, i.e. $m = v + b - 1$.

8.3.1 Characterization of robust designs

a) Designs with proportional frequencies

Designs with proportional frequencies are characterized by the property that $\mathbf{N} = \mathbf{rk}'/n$, or $n_{ij} = n_{i\cdot}\, n_{\cdot j}/n$, where $n_{i\cdot} = \sum_j n_{ij}$. In this case it is seen easily that $V(p_{ij})/\sigma^2 = 1/n_{i\cdot} + 1/n_{\cdot j} - 1/n$. Thus any design with proportional frequencies cannot be robust except in very special case where all n_{ij}'s are equal. From this, it follows that randomized complete block design is robust.

b) Balanced binary designs

We have seen in Chapter 3 that for a balanced design all the off diagonal elements of **A** are equal. Let the off diagonal elements of the matrix **A** of a binary, balanced design be $-\alpha$. Then $V(p_{ij})/\sigma^2 = \{1 + (n_{.j} - 1)/v\alpha\}/n_{\cdot j}$, which is a constant if and only if $n_{.j}$ is a constant for all j. Since the design is binary and balanced, this implies that $n_{i.}$ must also be a constant for all i. Hence a balanced binary design is robust if and only if it is equi-block sized and equi-replicate, i.e., if and only if it is a BIB design.

c) Partially balanced incomplete block design with two associate classes

Consider a two associate PBIB design with parameters $v, b, r, k, \lambda_i, n_i, p^i_{jk}$ $(i, j, k = 1, 2)$. Let D be the class of all 2-associate PBIB designs satisfying block structure:

(A): For any treatment i appearing in the j-th block, the number of first associates of i occuring in the same block is a constant (say g) independently of i and j.

It can be shown after some algebra that for a PBIB design belonging to D,

$$V(p_{ij})/\sigma^2 = 1/k + \{(k - 1) B_{22} + gB_{12}\}/\Delta,$$

where B_{22}, B_{12} and Δ are some constants dependent on the design parameters. Thus, all 2 associate PBIB designs, satisfying the block structure (A), are robust. The following are some examples of PBIB designs having the block structure (A). The value of g is also given for each case.

i) All non-group divisible two associate PBIB designs with $\lambda_2 = 0$, $g = k - 1$.

ii) All semi regular group divisible designs, $g = c - 1$, where $c = k/m$, there being $v = mn$ treatments divided into m groups of n treatments each.

iii) All triangular PBIB designs satisfying $r + (n - 4)\lambda_1 - (n - 3)\lambda_2 = 0$, $g = 4k/n - 2$, there being $v = n(n - 1)/2$ treatments.

iv) All latin square type designs with two constraints (L_2) satisfying $r + (s - 2)\lambda_1 - (s - 1)\lambda_2 = 0$, $g = 2k/s - 2$, there being $v = s^2$ treatments.

The term 'robustness' has been used in another way in the literature. Ghosh (1979) defined robustness in the following manner: A connected block design v, b, r, k is said to be robust against the unavailability of any t, a positive number $\leqslant r - 1$, observations, if the design obtained by omitting any t observations remains connected. Furthermore, Ghosh, Rao and Singhi (1983) characterized the robustness property of m-associate class of PBIB designs completely for $m = 2$ and partially for $m = 3$. Kageyama (1982) investigated the robustness of 3-associate PBIB designs based on various association schemes. The investigation is based on the condition for connectedness due to Ogawa, Ikeda and Kageyama (1982).

Consider a three associate PBIB design with parameters $v, b, r, k, \lambda_l, n_l, p^l_{jl}$ $(i, j, l = 0, 1, 2, 3)$. By the definition of connectedness, it is obvious that if all the λ_i's are positive then the design is connected. Ogawa, Ikeda and Kageyama (1982) presented the following results as a corollary of their main theorem:

Corollary 8.1 (a): A three associate PBIB design with $\lambda_1 = 0 < \lambda_2$, λ_3 is connected if and only if $p^1_{il} > 0$ for some integers $i, l \geqslant 2$.

Corollary 8 1 (b): A three associate PBIB design with $\lambda_1 = \lambda_2 = 0 < \lambda_3$ is connected if and only if $p^i_{33} > 0$ for all $i = 1, 2$, and then the latter condition $p^1_{23} > 0$ is superfluous.

Ghosh, Rao and Singhi (1983) characterized the robustness property of m-associate PBIB designs partially for $m = 3$. As a sufficient condition for the robustness of PBIB designs, their result essentially shows that if each of the association matrices $\mathbf{A}_1$, $\mathbf{A}_2$, $\mathbf{A}_3$ of an association scheme is connected, then a three associate PBIB design based on the association scheme is robust against the unavailability of any $t (\leqslant r - 1)$ observations. Thus, for showing the robustness of a three-associate PBIB design, it suffices to investigate the connectedness of association matrices.

In general, it holds that for any association scheme with three associate classes

$$\mathbf{A}_i \mathbf{A}_i' = n_i \mathbf{I}_v + \mathbf{p}^1_{ii} \mathbf{A}_1 + p^2_{ii} \mathbf{A}_2 + p^3_{ii} \mathbf{A}_3, \; i = 1, 2, 3,$$

which obviously implies that $\mathbf{A}_i$ itself is the incidence matrix of a symmetrical three associate PBIB design, based on the same association scheme as the original, with parameters $v = b$, $r = k = n_l$, $\lambda_j = p^j_{ii}$, $j = 1, 2, 3$, for each $i (= 1, 2, 3)$. In this sense, the connectedness of each $\mathbf{A}_i$ can be investigated as the connectedness of the incidence matrix of a symmetrical three associate PBIB design by using corollaries 8.1 (a) and (b). This observation is very powerful and has been used by Kageyama (1982) to study the robustness of three associate PBIB designs. The results obtained by Kageyama are briefly summarized below:

i) The robustness of a rectangular PBIB design cannot be established.
ii) A cubic PBIB design with $v = s^3$ treatments for $s \geqslant 3$ is robust against the unavailability of any $s(\leqslant r - 1)$ observations.
iii) An extended triangular PBIB design with $v = (s + 2)(s + 3)(s + 4)/6$ for $s \geqslant 3$ is robust against the unavailability of any $t(\leqslant r - 1)$ observations.
iv) The robustness of three associate group divisible design cannot be established.
v) The robustness of six series of three associate PBIB designs presented by Sprott (1955) cannot be established.

Readers interested in details of these results may refer to Kageyama (1982).

8.4 RESISTANT DESIGNS

In this Section we discuss another useful property of block designs known as resistance. The designs having the resistance property are called resistant designs. Susceptible designs are those which do not possess the property of resistance. The idea of resistance is due to Hedayat and John (1974).

Definition 8 3: A BIB design $D(v, b, r, k, \lambda)$ is called resistant of degree n if the design D^* obtained by deleting $n \leqslant v-2$ treatments from D is variance balanced. D is called globally resistant when any of the n treatments are deleted and is called locally resistant when D^* is variance balanced only when some particular set of n treatments are removed.

An extension of the above definition to more general block designs is fairly obvious. The resistance property may be useful in experimental situations requiring deletion of one or more treatments. Hedayat and John (1974) have listed several such situations. They have studied some characterizations of resistant designs and have given some methods of their construction. They have also indicated some open problems. We now discuss some of their results:

Let for a treatment x there exists a BIB design $D(v, b, r, k, \lambda)$ such that $D: D_x\ U D_{\bar{x}}$ where D_x consists of all the blocks which do not contain x and $D_{\bar{x}}$ is the set of blocks which contain x. Let $D_{\bar{x}'}$ be the design obtained by deleting the treatment x from $D_{\bar{x}}$. Then it is easy to prove the following theorem:

Theorem 8.1: D is locally resistant of degree one for x if and only if D_x is a BIB design.

The proof is fairly obvious by noting that D_x is a BIB design if and only if $D_{\bar{x}'}$ is also a BIB design. If D_x is a BIB design for all x then it follows that D is globally resistant of degree one.

Example 8.1: Consider the BIB design $D(8, 14, 7, 4, 3)$ with the block contents as

$$
\begin{array}{cccccccccccccc}
3 & 4 & 5 & 6 & 7 & 1 & 2 & 1 & 2 & 3 & 4 & 5 & 6 & 7 \\
5 & 6 & 7 & 1 & 2 & 3 & 4 & 2 & 3 & 4 & 5 & 6 & 7 & 1 \\
6 & 7 & 1 & 2 & 3 & 4 & 5 & 4 & 5 & 6 & 7 & 1 & 2 & 3 \\
7 & 1 & 2 & 3 & 4 & 5 & 6 & 8 & 8 & 8 & 8 & 8 & 8 & 8
\end{array}
$$

Here x is treatment number 8, the set of first seven blocks is D_x and the remaining seven blocks constitute $D_{\bar{x}}$. The design $D_{\bar{x}'}$ is BIBD and has the parameters (7, 7, 3, 3, 1) and D_x is BIBD with parameters (7, 7, 4, 4, 2). The design D is locally resistant of degree one if treatment 8 is deleted. It is also globally resistant of degree one as the resistance property is retained even if we delete any treatment other than 8.

Hedayat and John (1974) showed through examples that the property of resistance depends not only on the parameters of the design but also on the way the design is constructed. It is also clear from the theorem 8.1 and subsequent arguments that in D every triplet containing x appears the same number of times if D is locally resistant and for a globally resistant design, D every triplet of treatments must appear a constant number of times. This leads to the characterization that every globally resistant design of degree one must be a 3-design, i.e. it is doubly balanced. It immediately follows that a t-design is globally resistant of degree t-2.

Hedayat and John further studied some characterizations and construction problems for resistant designs of degree one. Other relevant references on the topic are of Most (1975), Shah and Gujarati (1978, 1983), Chandak (1980) and Kageyama (1982). These papers generalize the results of Hedayat and John in several directions. Chandak (1980) and Kageyama (1982) consider the resistance property of a general balanced block design and report several interesting results.

8.5 GRUBB'S ESTIMATORS IN BLOCK DESIGN

Consider the general block design $D(v, b, \mathbf{r}, \mathbf{k})$ and the usual intra block model with a slight modification that $E(e_{ijk}^2) = \sigma_j^2$, $j = 1,\ldots, b$, a departure from the usual assumption of homogeneity of error variances.

In a block design, the entire experimental material assumed to be heterogeneous, is divided into blocks so that the experimental units within a block are as homogeneous as possible and the between blocks variation is taken away from the experimental error, resulting thereby into more precise estimation of treatments contrasts. In view of this it is not unrealistic to assume that the errors or the observations arising from a given block have a constant variance, but errors arising from different blocks have different variances. A fairly extensive bibliography on the variance estimators initiated by Grubb's (1948) initial paper on the subject is summarized in Maloney (1973). Russell and Bradley (1958) presented two tests of variance homogeneity, an exact test on the assumption that the first $(b-1)$ variances are homogeneous and an approximate test when $b = 3$. Ellenberg (1977) obtained the joint distribution of $S_1,\ldots, S_k$, $k \leqslant b$, where S_j is the sum of squared residuals for block j, under the homogeneity assumption, but regarded the complicated series expressions obtained as interactable. Brindley and Bradley (1985) obtained the small sample joint distribution of S_1, S_2 and S_3 when $b = 3$ and the distribution of a variance–homogeneity test statistic based on them in simple form, both under the assumption of homogeneous variances and the general alternative to the assumption. They gave an approximate procedure for $b > 3$ and also obtained the power function of the exact test of Russell and Bradley. Bradley and Rollier (1985) obtained extensions of Grubb's estimators for classes of incomplete block designs and considered inference procedures based on them to the extent possible.

8.5.1 Estimators for general block designs

Define

$$S_j = \sum_{i=1}^{v} \sum_{k=1}^{n_{ij}} (y_{ijk} - \hat{\mu} - \hat{\tau}_i - \hat{\beta}_j)^2, \; j = 1, \ldots, b. \tag{8.3}$$

Substituting for $\hat{\mu}$, $\hat{\tau}_i$, $\hat{\beta}_j$, we have

$$\begin{aligned}(y_{ijk} - \hat{\mu} - \hat{\tau}_i - \hat{\beta}_j) = e_{ijk} &- \sum_s \sum_t \sum_k a^-_{is}\, e_{stk} \\ &+ \sum_h \sum_s \sum_t \sum_k \frac{n_{hj}}{k_j} a^-_{hs}\, e_{stk} \\ &+ \sum_h \sum_s \sum_t \sum_k \frac{n_{ht}}{k_t} a^-_{ih}\, e_{stk} \\ &- \sum_h \sum_s \sum_t \sum_u \sum_k \frac{n_{hj} n_{ut}}{k_j k_t} a^-_{hu}\, e_{stk} - \sum_s \sum_k \frac{e_{sjk}}{k_j},\end{aligned} \tag{8.4}$$

where a^-_{hs} is the (h, s)-th element of $\mathbf{A}^-$. Using the fact that $E(e^2_{ijk}) = \sigma^2_j$, we have

$$\begin{aligned}E(S_j) = \sigma_j^2 \Big[k_j - 1 - 2 \sum_s n_{sj}\, a^-_{ss} + 2 \sum_h \sum_s \frac{n_{sj} n_{hj}}{k_j} a^-_{hs} \Big] \\ + \sum_t \sigma_t^2 \Big[\sum_h \sum_s n_{ht} n_{sj}\, (a^-_{hs})^2 + \Big\{\frac{1}{k_j k_t}\Big\} \\ \{ \sum_h \sum_s n_{hj} n_{st} a^-_{hs} \}^2 - \sum_h \frac{n_{ht}}{k_j} \{ \sum_s n_{sj}\, a^-_{hs} \}^2 \\ - \sum_h \frac{n_{hj}}{k_t} \{ \sum_s n_{st} a^-_{hs} \}^2 \Big].\end{aligned} \tag{8.5}$$

For randomized complete block design

$$E(S_j) = \frac{(v-1)(b-2)}{b} \sigma_j^2 + \frac{(v-1)}{b^2} \sum_t \sigma_t^2 \tag{8.6}$$

We also have the error sum of squares for the design, $R_0^2 = \sum_j S_j$. From (8.6), we have

$$E(R_0^2) = [(v-1)(b-1)/b] \sum_t \sigma_t^2$$

and that

$$G_j = \frac{b}{(v-1)(b-2)} S_j - \frac{1}{(v-1)(b-1)(b-2)} R_0^2 \qquad (8.7)$$

is an unbiased estimator of $\sigma_j^2 \cdot G_j$ is Grubb's estimator.

For a general block design, let $\boldsymbol{\sigma}^2 = (\sigma_1^2,\ldots, \sigma_b^2)'$, $\mathbf{S} = (S_1,\ldots, S_b)'$ and $\mathbf{G} = (G_1,\ldots, G_b)'$. Then

$$E(\mathbf{S}) = \mathbf{Z}\boldsymbol{\sigma}^2 \qquad (8.8)$$

and

$$\mathbf{G} = \mathbf{Z}^{-1}\mathbf{S}, \qquad (8.9)$$

thus defining the extension of Grubb's estimators. Elements of $\mathbf{Z}$ in (8.8) are determined by (8.5) and the non-singularity of $\mathbf{Z}$ is assumed.

Special forms of (8.5) and $\mathbf{Z}$ in (8.8) have been examined for well known classes of incomplete block designs by Bradley and Rollier (1985). For PBIB designs with m associate classes the expressions for $E(S_j)$ and the matrix $\mathbf{Z}$ have been computed. These expressions are quite lengthy. For two associate PBIB designs, there are no major simplifications in these expressions. Sufficient simplifications, however, occur in $\mathbf{Z}$ for semi regular GD and singular GD designs to demonstrate the existence of $\mathbf{Z}^{-1}$ for designs of practical use in these classes. For $\lambda_1 = 0$ or $\lambda_1 \neq 0$, $r \leqslant 10$, the existence of $\mathbf{Z}^{-1}$ has been demonstrated for the semi regular GD designs showing thereby the existence of Grubb's estimators for these designs. The existence of the estimators G_j for σ_j^2 has been established for all singular GD designs when the semi regular GD design is λ times linked and when the singular GD design is $n\lambda_2$ times linked, explicit forms of G_j are obtained. Similarly, explicit expressions have been obtained for two replicate, singly linked, triangular PBIB(2) designs.

For BIB designs $\mathbf{Z}$ is most easily obtained from (8.5). For $\lambda v > 2k$, $\mathbf{Z}^{-1}$ exists and $\mathbf{Z}$ is positive definite. For the dual of a symmetrical BIB design the Grubb's estimators always exist. Other more specialized designs with specific estimators considered are BIB designs with PBIB m associate class and triangular, singly linked block duals.

8.5.2 Generalized test procedures

The first test of Russell and Bradley (1958) for testing H_0: $\sigma_1^2 = \sigma^2$, against H_a: $\sigma_1^2 \neq \sigma^2$ given $\sigma_j^2 = \sigma^2$, $j = 2,\ldots, b$, has been extended by Bradley and Rollier (1985) to all BIB designs and to PBIB (m) designs for which all pairs of treatments in block one are u-th associates. The test is based on a Snedecor's F distribution as is its power function.

Consider a connected, binary block design. Without loss of generality, attention is focussed on block 1. We wish to test the null hypothesis $H_0: \sigma_1^2 = \sigma^2$, against the alternative hypothesis, $H_a: \sigma_1^2 \neq \sigma^2$, given that $\sigma_j^2 = \sigma^2$, $j = 2,\ldots, b$. The test compares S_1 with $R_0^2 - \alpha S_1$, where α is a scalar constant to be determined. Under H_0: $E(R_0^2) = (n-v-b+1)\sigma^2$. For a randomized complete block design, the test statistic for testing H_0 against H_a is

$$F = \frac{\alpha(b-2)S_1}{(R_0^2 - \alpha S_1)} = \frac{b(b-2)S_1}{(b-1)R_0^2 - bS_1}$$

Under H_0, F has the Snedecor's F-distribution with $(v-1)$ and $(b-2)(v-1)$ degrees of freedom. This result was obtained earlier by Russel and Bradley (1958). Under H_a

$$\frac{\alpha(b-2)\sigma^2 S_1}{[\sigma_1^2 + (\sigma^2 - \sigma_1^2)/b][R_0^2 - \alpha S_1]} = \frac{bF}{1+\gamma_1}$$

has the Snedecor's F distribution with $(v-1)$ and $(b-2)(v-1)$ degrees of freedom, where $\gamma_1 = (b-1)\sigma_1^2/\sigma^2$. This result has also been obtained by Brindley and Bradley (1985). For a BIB design, the test statistic is

$$F = \lambda v\,[(b-1)(k-1) - (v-1)]S_1/(k-1)[(\lambda v - k)R_0^2 - \lambda v S_1]$$

which has the Snedecor's F distribution with $(k-1)$ and $[(b-1)(k-1) - (v-1)]$ degrees of freedom under H_0 and

$$(\lambda v/k)F/(1+\gamma_1)$$

has the same F distribution under H_a with

$$\gamma_1 = (\lambda v - k)\sigma_1^2/k\sigma^2.$$

Consider any PBIB design with m associate classes such that any pair of treatments in block 1 are u-th associates, $u = 1,\ldots,m$. For such designs the statistic

$$F = \alpha\,[(b-1)(k-1) - (v-1)]S_1/(k-1)(R_0^2 - \alpha S_1)$$

and

$$F/(a - a^*)\,(1+\gamma_1)$$

have the Snedecor's F distribution with $(k-1)$ and $[(b-1)(k-1) - (v-1)]$ degrees of freedom under H_0 and H_a respectively with $\gamma_1 = (1 - a + a^*)\sigma_1^2/(a - a^*)\sigma^2$. Here a and a^* are the elements of $\mathbf{A}^-$ for block 1 such that $a_{ii'}^-$ is a if $i = i'$ and a^* if $i \neq i'$.

Useful designs belonging to this class are two associate class PBIB designs with either λ_1 or λ_2 zero.

We now give some general results of which the cases just described are special cases.

Let

$$S_j = \sum_{i=1}^{v} n_{ij} r_{ij}^2\,,\; j = 1,\ldots,b. \tag{8.10}$$

and define

$$b_{st} = \Sigma_h \, n_{ht} a_{hs}^- / k_t$$
$$\varepsilon_{tt'} = \sum_h \sum_s n_{ht} n_{st'} \, a_{hs}^- / k_t k_{t'}, \; s = 1, \ldots, v, \qquad (8.11)$$
$$t, t' = 1, \ldots, b.$$

From the expression (8.4),

$$n_{ij} r_{ij} = n_{ij}(e_{ij} - \sum_s n_{sj} e_{sj}/k_j - \sum_s \sum_t n_{st} a_{is}^- e_{st} + \sum_s \sum_t n_{st} b_{it} e_{st}$$
$$+ \sum_s \sum_t n_{st} b_{sj} \; e_{st} - \sum_s \sum_t n_{st} \varepsilon_{jt} e_{st}) \qquad (8.12)$$

The expression (8.12) may be written with the e_{st}'s replaced by y_{st}'s and $n_{ij} r_{ij}$ is a linear function of orthogonal error contrasts. In addition, $\Sigma \, n_{ij} \, r_{ij} = 0$. The covariance matrix of $n_{ij} \, r_{ij}$, $i = 1, \ldots, v$, for each j under H_a is

$$Cov \; (n_{ij} r_{ij}, n_{i'j} r_{i'j}) = n_{ij} n_{i'j} \; \{[\delta_{1j} \sigma_1^2 + (1 - \delta_{1j})\sigma^2]$$
$$(\delta_{ii'} - a_{ii'}^- - 1/k_j + b_{ij} + b_{i'j} - \varepsilon_{jj})$$
$$+ (\sigma_1^2 - \sigma^2)[\sum_s n_{s1} a_{is}^- \, a_{i's}^- + k_1 b_{i1} b_{i'1} \qquad (8.13)$$
$$- \sum_s n_{s1} b_{sj} a_{is}^- - \sum_s n_{s1} b_{sj} a_{i's}^- + k_1 b_{i1} \, \varepsilon_{1j} + k_1 b_{i'1} \varepsilon_{1j}$$
$$+ \sum_s n_{s1} b_{sj}^2 - k_1 \, \varepsilon_{ij}^2 - \delta_{1j}(a_{ii'}^- - b_{i1} - b_{i'1} + \varepsilon_{11})]\}$$

Since $\quad \mathbf{A^- A A^-} = \mathbf{A^-}$ and $\mathbf{A} = \mathbf{R} - \mathbf{N K^{-1} N'}$,

$$\sum_i r_i a_{hi}^- a_{sl}^- = a_{hs}^- + \sum_j k_j b_{hj} b_{sj}. \qquad (8.14)$$

Repeated use of (8.14) is required to obtain (8.13) Since $\mathbf{AA^-}$ is idempotent and $R(\mathbf{AA^-}) = R(\mathbf{A})$, trace $(\mathbf{AA^-}) = R(\mathbf{AA^-}) = R(\mathbf{A}) = v - 1$, for a connected design,

$$\text{trace } (\mathbf{AA^-}) = R(\mathbf{A}) = (v - 1) = \sum_i r_i \, a_{ii}^- - \sum_j k_j \, \varepsilon_{jj},$$

and this is enough to verify that

$$E(R_0^2) = \sum_i \sum_j Cov \; (n_{ij} r_{ij}, \, n_{ij} r_{ij}) = (n - v - b + 1)\sigma^2 \text{ under } H_0.$$

The association scheme of the dual design enters only in determining $\varepsilon_{jj'}$, $j \neq j'$, in (8.12) and only ε_{11} is required for the special classes of designs considered above.

We have the association matrices

$\mathbf{B}_u = (b^u_{hs})$, where $b^u_{hs} = 1$ or 0 as treatments h and s are or are not u-th associates, $u = 1,\ldots, m$, and $\delta_{vw} = 1$ or 0 as $v = w$ or $v \neq w$.

References

Addelman, S. (1962). Orthogonal main-effect plans for asymmetrical factorial experiments. Technometrics, 4, 21–46.

Atiquallah, M. (1961). On a property of balanced designs. Biometrika, 48, 215–218.

Bartlett, M.S. (1978). An introduction to stochastic processes (3rd ed). Cambridge University Press, Cambridge.

Bhattacharya, K.N. (1945). On a new symmetrical balanced incomplete block design. Bulletin of Calcutta Mathematical Society, 36, 91–96.

Boopathy, G.N. (1980). Some contributions to designs and an analysis of bioassays. Unpublished Ph.D. thesis, IARI, New Delhi.

Bose, R.C. (1939). On the construction of BIB designs. Annals of Eugenics, 9, 353–399.

Bose, R.C. (1947a). Presidential address. Proceedings of the 34-th Indian Science Congress, p. 12.

Bose, R.C. (1947b). Mathematical theory of the symmetrical factorial designs. Sankhya, 8, 107–166.

Bose, R.C. (1963). Strongly regular graphs, partial geometries and partially balanced designs. Pacific Journal of Mathematics, 13, 389–419.

Bose, R.C., Clatworthy, W.H. and Shrikhande, S.S. (1954). Tables of PBIB designs with two associate classes. North Carolina Agricultural Experimental Bulletin, No. 107.

Bose, R.C. and Connor, W.S. (1952). Combinatorial properties of group divisible incomplete block designs. Annals of Mathematical Statistics, 23, 367–383.

Bose, R.C. and Mesner, D.M. (1959). On linear associative algebra corresponding to association scheme of PBIB designs. Annals of Mathematical Statistics, 30, 21–38.

Bose, R.C. and Nair, K.R. (1939). PBIB designs. Sankhya, 4, 337–372.

Bose, R.C. and Shimamoto, T. (1952). Classification and analysis of PBIB designs with two associate classes. Journal of the American Statistical Association, 47, 151–184.

Bose, R.C. and Shrikhande, S.S. (1960). On the composition of BIB designs. Canadian Journal of Mathematics, 12, 177–188.

Box, G.E P. and Draper, N.R. (1975). Robust designs. Biometrika, 62, 347–352.

Bradley, R.A. and Rollier, D A. (1985). The extension of Grubb's estimators for classes of incomplete block designs. Technical Report No. STA 14, Department of Statistics and Computer Sciences, University of Georgia, Athens, GA 30602, USA.

Bradley, R.A., Walpole, R.E. and Kramer, C.Y. (1960). Intra and inter block analysis for factorials in incomplete block designs. Biometrics, 16, 566–581.

Bradley, R.A. and Yeh, C.M. (1980) Trend free block designs. Annals of Statistics, 8, 883-893.

Brenna, L.S. (1958). Factorial treatments in lattice designs. Ph.D. dissertation, Virginia Polytechnic Institute Library, Blacksburg, Virginia.

Brenna, L.S. and Kramer, C.Y. (1961). Factorial treatments in rectangular lattice designs. Journal of the American Statistical Association, 56, 368–378.

Brindley, D.A. and Bradley, R.A. (1985). Some new results on Grubb's estimators. Journal of the American Statistical Association, 80, 711–714.

Calinski, T. (1971). On some desirable patterns in block designs. Biometrics, 27, 275–292.

Calvin, L.D. (1954). Doubly balanced incomplete block designs for experiments in which the treatments are correlated. Biometrics, 10, 61–83.

Cameron, J.M. and Speckman, J.A. (1972). Use of cyclic designs, designs in calibration. Appendix to cyclic designs. National Bureau of Standards, Applied Mathematics, Series 62, 12-15, Washington.

Ceranka, B. and Mejza, S. (1978). On the efficiency factor for a contrast of treatment parameters. Biometrical Journal, 21, 99–102.

Chakrabarti, M.C. (1963). On the use of incidence matrices of designs in sampling from finite populations. Journal of the Indian Statistical Association, 1, 78–85.

Chandak, M.L. (1980). On the theory of resistant block designs. Calcutta Statistical Association Bulletin, 29, 27–34.

Cheng, C.S. (1978). Optimality of a certain asymmetrical experimental design. Annals of Statistics, 6, 1239-1261.

Cheng, C.S. (1979). Optimal incomplete block designs with four verieties. Sankhya, B, 41, 1–14.

Cheng, C.S. (1980). On the E-optimality of some block designs. Journal of the Royal Statistical Society, B, 42, 199–204.

Cheng, C.S. and Wu, C.F. (1981). Nearly balanced incomplete block designs. Biometrika, 68, 493-500.

Clatworthy, W.H. (1956). Contributions on partially balanced incomplete block designs with two associate classes. National Bureau of Standards, Applied Mathematics Series No. 47, Washington, D.C.

Clatworthy, W.H. (1973). Tables of two associate classes partially balanced designs. U.S. Department of Commerce, National Bureau of Standards, Washington, D.C.

Conniffe, D. and Stone, J. (1974). The efficiency factor of a class of incomplete block design. Biometrika, 61, 633–634.

Conniffe, D. and Stone, J. (1975). Some incomplete block designs of maximum efficiency. Biometrika, 62, 685–686.

Connor, W.S. and Clatworthy, W.H. (1954). Some theorem for partially balanced designs. Annals of Mathematical Statistics, 85, 100–112.

Cornish, F.A. (1938). Factorial treatments in incomplete randomized block. Journal of the Australian Institute of Agricultural Sciences, 4, 199–203.

Cornish, F.A. (1940). The analysis of quasi-factorial designs with incomplete data. Journal of the Australian Institute of Agricultural Sciences, 6, No. 1, 31–39.

Corseen, L.C.A. (1958). Vectors, a tool in statistical regression theory. Mededelinger Van de Landbouwhogeschoolte wageningen/Nederland, 58, 1–92.

Cotter, S.C., John, J.A. and Smith, T.M F. (1973). Multifactor experiments in non-orthogonal designs. Journal of the Royal Statistical Society, B, 35, 361–367.

Curnow, R.N. (1963). Sampling the diallel cross. Biometrics, 19, 287–306.

Das, M.N. (1958). On reinforced incomplete block designs. Journal of the Indian Society of Agricultural Statistics, 10, 73–77.

Das, M.N. (1960a). Factorial replicates as asymmetrical factorial designs. Journal of the Indian Society of Agricultural Statistics, 11, 73–110.

Das, M.N. (1960b). Circular designs. Journal of the Indian Society of Agricultural Statistics, 12, 46–56.

Das, M.N. (1980). Presidential address delivered at the Indian Science Congress Association, 68-th Session, Section of Statistics.

Das, M.N. and Ghosh, D.K. (1985). Balancing incomplete block designs. Sankhya, B, 47, 67–77.

Das, M.N. and Kulkarni, G.A. (1966). Incomplete block designs for bio-assays. Biometrics, 22, 706–729.

Das, M.N. and Rao, S V.S.P. (1968). On balanced n-ary designs. Journal of the Indian Statistical Association, 6, 137–146.

Das, M.N. and Sivaram, K. (1968). Partial diallel crosses and incomplete block designs. Contribution in Statistics and Agricultural Sciences. Volume presented to Dr V.G. Panse at his 62nd birthday, 49-59.

David, H.A. (1963). The structure of cyclic paired-comparison designs. Journal of the Australian Mathematical Society, 3, 117–127.

David, H.A. (1965). Enumeration of the cyclic paired comparison designs. American Mathematical monthly, 72, 241–248.

David, H.A. (1967). Resolvable cyclic designs. Sankhya, 29, 191–198.

David, H.A. and Wolock, F.W. (1965). Cyclic designs. Annals of Mathematical Statistics, 36, 1526–1534.

Dean, A.M. and John, J.A. (1975). Single replicate factorial experiments in generalized cyclic designs-II: Asymmetrical arrangements. Journal of the Royal Statistical Society, B, 37, 72–76.

Dean, A.M. and Lewis, S.M. (1980). A unified theory for generalized cyclic designs. Journal of Statistical Planning and Inference, 4, 13–23.

Dey, A. (1970). On construction of balanced n-ary block designs. Annals of the Institute of the Statistical Mathematics, 22, 389–393

Dey, A. (1975). A note on balanced designs. Sankhya, B, 37, 461–462.

Dey, A. and Gupta, V.K. (1986). Another look at the efficiency and partially efficiency balanced designs. Sankhya, B, 48, 437–438.

Dey, A., Singh, M. and Saha, G.M. (1981). Efficiency balanced block designs. Communications in Statistics, A, 10(3), 237–247.

Eccleston, J.A. and Hedayat, A. (1974). On the theory of connected designs: Characterization and optimality. Annals of Statistics, 2, 1238–1255.

Eisenhart, C. (1947). The assumptions underlying the analysis of variance. Biometrics, 3, 1–21

Ellenberg, J.H. (1977) The joint distribution of the standardized row sums of squares from a balanced two-way layout. Journal of the American Statistical Association, 72, 407–411.

Federer, W.T. (1955). Experimental designs--theory and applications. Macmillan, New York.

Federer, W.T. (1956). Augmented (or hoonuiaku) designs. Hawaiian planters. Record, 55, 191-208.

Federer, W.T. (1961). Augmented designs with one-way elimination of heterogeneity. Biometrics, 17, 447-473.

Federer, W.T. and Raghavarao, D. (1975). On augmented designs. Biometrics, 31, 29–35.

Finney, D.J. (1978). Statistical methods in biological assay. 3rd edition. Charles Griffen and Company Ltd., U.K.

Fisher, R.A. (1940). An examination of the different possible solutions of a problem in incomplete blocks. Annals of Eugenics, 10, 52–75.

Fisher, R.A. and Yates, F. (1963). Statistical tables for biological, agricultural and medical research. Longman, U.K.

Foody, W. and Hedayat, A. (1977). On theory and applications of BIB designs with repeated blocks. Annals of Statistics, 5, 932–945.

Fyfe, J.L. and Gilbert, N.E.G. (1963). Partial diallel cross. Biometrics, 19, 276–286.

Ghosh, S. (1979). On robustness of designs against incomplete data. Sankhya, B, 40, 204–208.

Ghosh, S., Rao, S.B. and Singhi, N.M. (1983). On a robust property of PBIB designs. Journal of Statistical Planning and Inference, 8, 355–364.

Gopalan, R. and Dey, A. (1976). On robust experimental designs. Sankhya, B, 38, 297–299.

Grubbs, F.E. (1948). On estimating precision of measuring instruments and product variability. Journal of the American Statistical Association, 43, 243–264.

Gupta, S.C. (1983). A basic lemma and the analysis of block and kronecker product designs. Journal of Statistical Planning and Inference, 7, 407–416.

Gupta, V.K., Nigam, A.K. and Kumar, P. (1982). On a family of sampling schemes with inclusion probability proportional to size. Biometrika, 60, 191–196.

Gupta, V.K., Nigam, A.K. and Puri, P.D. (1987). Characterization and construction of incomplete block designs for symmetrical parallel line assay. Journal of the Indian Society of Agricultural Statistics, 34, No. 2, 161–166.

Hall, M. Jr. (1967). Combinatorial theory. Blaisdell Publishing Company. London.

Harshbarger, B. (1954). The 2^3 factorial in a latinized rectangular lattice design. Technical report No. 1, Virginia Agricultural Experimental Station.

Hedayat, A. and Federer, W.T. (1974). Pair-wise and variance balanced incomplete block designs. Annals of the Institute of the Statistical Mathematics, 26, 331–338.

Hedayat, A. and John, P.W.M. (1974). Resistant and susceptible BIB designs. Annals of Statistics, 2, 148–158.

Hedayat, A. and Kegeyama, S. (1980). The family of t-designs part I. Journal of Statistical Planning and Inference, 4, 173–212.

Hinkelman, K. (1960). Kruenzungsplane Zur selektion c zuchteing bei Waldbaumen. Silvae Genetica, 9, 121–133.

Hinkelman, K. (1964). Extended group divisible partially balanced incomplete block designs. Annals of Mathematical Statistics, 35, 681–695.

Hinkelman, K. (1965). Unvollstandige Triallele, Biometrische Zeitschrift, 8, 331-349.

Hinkelman, K. (1966). Unvollistandige diallele Kruzungsplane. Biometrische Zeitschrift, 8, 242–265.

Hinkelman, K. and Kempthorne, O. (1963). Two classes of group divisible partial diallel crosses. Biometrika, 50, 281–291.

Hoblyn, T.N., Pearce, S.C. and Freeman, G.H. (1954). Some considerations in the designs of successive experiments in fruit plantations. Biometrics, 13, 503–515.

Hoffman, A.J. (1963). On the duals of symmetric partially balanced incomplete block designs. Annals of Mathematical Statistics, 34, 528–531.

Jacroux, M.A. (1978). On the properties of proper (M-S) optimal block designs. Annals of Statistics, 6, 1302–1309.

James, A.T. and Wilkinson, G.N. (1971). Factorization of the residual operator and canonical decomposition of non-orthogonal factors in the analysis of variance. Biometrika, 58, 279-294.

John, J.A. (1965). A note on the analysis of incomplete block experiments. Biometrika, 52, 633-636.

John, J.A. (1966). Cyclic incomplete block designs. Journal of the Royal Statistical Society, B, 28, 345–360.

John, J.A. (1969) A relationship between cyclic and PBIB designs. Sankhya, B, 31, 535–540.

John, J.A. (1973). Generalized cyclic designs in factorial experiments. London: Collier-Macmillan.

John, J.A. (1978). A note on a finite population plan. Annals of Statistics, 6, 697–699.

John, J.A. (1981). Factorial balance and the analysis of design with factorial structure. Journal of Statistical Planning and Inference, 5, 99–105.

John, J.A. and Dean, A.M (1975). Single replicate factorial experiments in generalized cyclic designs: I, symmetrical arrangements. Journal of the Royal Statistical Society, B, 37, 63–71.

John, J.A. and Mitchell, T.J. (1977). Optimal incomplete block designs. Journal of the Royal Statistical Society, B, 39, 39-43.

John, J.A. and Smith, T.M.F. (1972). Two factor experiments in non-orthogonal designs. Journal of the Royal Statistical Society, B, 34, 401-409.

John, J.A., Wolock, F.W. and David, H.A. (1972). Cyclic designs. National Bureau of Standards, Applied Mathematics Series, 62, Washington, D.C.

John, P.W.M. (1964). Balanced designs with unequal number of replicates. Annals of the Mathematical Statistics, 35, 897–899.

John, P.W.M (1966). An extension of the triangular association scheme to three associate classes. Journal of the Royal Statistical Society, B, 28, 361–365.

John, P.W.M. (1971). Statistical design and analysis of experiments. London: Collier-Macmillan.

Jones, R.M. (1959). On a property of incomplete blocks. Journal of the Royal Statistical Society, B, 21, 172–179.

Kageyama, S. (1972) On the reduction of associate classes for certain PBIB designs. Annals of Mathematical Statistics, 43, 1528–1540.

Kageyama, S. (1974). Reduction of associate classes for block designs and related combinatorial arrangements. Hiroshima Mathematical Journal, 4, 527–618.

Kageyama, S. (1976). Resolvability of block designs. Annals of Statistics, 4, 655–661.

Kageyama, S. (1980). On properties of efficiency balanced designs. Communications in Statistics, A 9(6), 597-616.

Kageyama, S. (1981). Construction of efficiency-balanced designs. Communications in Statistics, A 10(6), 559-580.

Kageyama, S. (1982). The existence of locally resistant BIB designs of degree one. Technical Report No. 18182, Stat. Math. Division, Indian Statistical Institute.

Kageyama, S. and Hedayat, A. (1983). The family of *t*-designs-Part II. Journal of Statistical Planning and Inference, 7, 257-288.

Kageyama, S. and Tsuji, T. (1980). Some bounds on balanced block designs. Journal of Statistical Planning and Inference, 4, 155–167.

Kempthorne, O. (1952). The design and analysis of experiments. Wiley, New York.

Kempthorne. O. (1953). A class of experimental designs using blocks of two plots. Annals of Mathematical Statistics, 24, 76–84.

Kempthorne, O. and Curnow, R.N. (1961). The partial diallel cross. Biometrics, 17, 229–250, correction, 18, 128.

Khatri, C.G. (1982). A note on variance balanced designs. Journal of Statistical Planning and Inference, 6, 173–177.

Kiefer, J. (1958). On the non-randomized optimality and randomized non-optimality of symmetrical designs. Annals of Mathematical Statistics, 29, 675–699.

Kiefer, J. (1959). Optimum experimental designs. Journal of the Royal Statistical Society, B, 21, 272–319.

Kiefer, J. (1975). Construction and optimality of generalized Youden designs. In a survey of Statistical Designs and Linear Models (J.N. Srivastava), North Holland, Amsterdam, 333–353.

Kramer, C.Y. and Bradley, R.A. (1957a). Intra block analysis of factorials in two associate class group divisible designs. Annals of Mathematical Statistics, 28, 349–361.

Kramer, C.Y. and Bradley, R.A. (1957b). Examples of intra-block analysis for factorials in group divisible, partially balanced incomplete block designs. Biometrics, 13, 197–224.

Kshirsagar, A.M. (1958). A note on incomplete block designs. Annals of Mathematical Statistics, 29, 910–914.

Kshirsagar, A.M. (1966). Balanced factorial designs. Journal of the Royal Statistical Society, B, 28, 559–567.

Kulshrestha, A.C. (1971). Incomplete block designs and their application to bio-assays. Unpublished Ph. D. thesis. Institute of Advance Studies. Meerut University, India.

Kulshrestha, A.C., Dey, A. and Saha, G.M. (1972). Balanced designs with unequal replications and unequal block sizes. Annals of Mathematical Statistics, 43, 1342–1345.

Kumar P., Gupta, V.K. and Nigam, A.K. (1985). On inclusion probability proportional to size scheme. Journal of Statistical Planning and Inference, 12, 127–131.

Kurkjian, B. and Zelen, M. (1962). A calculus for factorial arrangements. Annals of Mathematical Statistics, 33, 600–619.

Kurkjian, B. and Zelen, M. (1963). Application of calculus of factorial arrangements I; block and direct product designs. Biometrika, 50, 63–73.

Kusumoto, K. (1965). Hyper cubic designs. Wakayama Medical Reports, 9, 123–132.

Kusumoto, K. (1967). Association schemes of new types and necessary conditions for the existence of regular and symmetrical PBIB designs with those association schemes. Annals of the Institute of Statistical Mathematics, 19, 73–100.

Kyiwin and Dey, A. (1980). Incomplete block designs for parallel line assays. Biometrics, 36, 487–492.

Maloney, C.J. (1973). Grubb's estimators to date. Proceedings of the 18-th conference on design of experiments in army research, development and testing, Part 2, 523–547.

McKeon, J.J. (1960). Some cyclical incomplete paired comparison designs. Technical Report No. 24, Psychometric laboratory of University of North Carolina.

Mirsky, L. (1972). An introduction to linear algebra. Clarendon press, Oxford.

Morgan, E. (1977). Construction of ternary designs. Utilitas Mathematica, 11, 3–31.

Most, B.M. (1975). Resistance of balanced incomplete block designs. Annals of Statistics, 3, 1149–1162.

Mote, V.L. (1958). On a minimax property of a balanced incomplete block design. Annals of Mathematical Statistics, 29, 910–914.

Mukerjee, R. (1979). Inter-effect orthogonality in factorial experiments. Calcutta Statistical Association Bulletin, 28, 83–108.

Mukerjee, R. (1980). Further results on the analysis of factorial experiments. Calcutta Statistical Association Bulletin, 29, 1–26.

Mukerjee, R. (1981). Construction of effect-wise orthogonal factorial designs. Journal of Statistical Planning and Inference, 5, 221–229.

Murty, J.S. and Das, M.N. (1967). Balanced n-ary block designs and their uses. Journal of the Indian Statistical Association, 5, 73–82.

Nair, K.R. and Rao, C.R. (1942). A note on partially balanced incomplete block designs. Science and Culture, 7, 568–569.

Nam, N.K. (1983). Computer aided construction of optimal designs. Unpublished Ph.D. thesis, Indian Agricultural Research Institute, New Delhi.

Neudecker, H. (1969). Some theorems on matrix differentiation with special reference to Kronecker matrix product. Journal of the American Statistical Association, 64, 953-963.

Nigam, A.K. (1974). Construction of balanced n-ary block designs and partially balanced arrays. Journal of the Indian Society of Agricultural Statistics, 26, No. 2, 48–56.

Nigam, A.K. (1976). Nearly balanced incomplete block designs. Sankhya, B, 38, 195–198.

Nigam, A.K. and Boopathy, G.M. (1985). Incomplete block designs for symmetrical parallel line assay. Journal of Statistical Planning and Inference, 11, 111–117.

Nigam, A.K. and Gupta, V.K. (1984). A method of sampling with equal or unequal probabilities without replacement. Applied Statistics, 33, 227-229.

Nigam, A.K., Gupta, V.K. and Narain, P. (1981). Monograph on supplemented block designs. Indian Agricultural Statistics Research Institute, New Delhi-12.

Nigam, A.K., Kumar, P. and Gupta, V.K. (1984). Some methods of inclusion probability proportional to size sampling. Journal of the Royal Statistical Society, B, 47, 564–571.

Nigam, A K. and Mathur, S.N. (1974). Kronecker sum of matrices and Kronecker product association schemes. Technical Report, Indian Agricultural Statistics Research Institute, New Delhi.

Nigam, A.K. and Puri, P.D. (1982). On partially efficiency balanced designs–II. Communication in Statistics, Theory and Methods, 11, 2817-2830.

Ogawa, J., Ikeda, S. and Kageyama, S. (1982). Connectedness of PBIB designs with applications. Paper presented at the seminar on combinatorials and applications in honour of Prof. S.S. Shrikhande.

Paik, U.B., and Federer, W.T. (1973). Partially balanced designs and property A and B. Communications in Statistics, Theory and Methods, 1, 331–350.

Pal, S. (1980). A note on partially efficiency balanced designs. Calcutta Statistical Association Bulletin, 29, 185–190.

Patterson, H.D. and Williams, E.R. (1976a). A new class of resolvable incomplete block designs. Biometrika, 63, 83–92.

Patterson, H.D. and Williams, E.R. (1976b). Some theoretical results on general block designs. In proceeding of the fifth British combinatorial conference Numerantium XV, 489–496, Utilitas Mathematica, Winnipeg.

Pearce, S.C. (1960). Supplemented balance. Biometrika, 47, 263-271.

Pearce, S.C. (1963). The use and classification of non orthogonal designs (with discussion). Journal of the Royal Statistical Society, 126, 353–377.

Pearce, S.C. (1964). Experimenting with blocks of natural sizes. Biometrics, 20, 699-706.

Pearce, S.C. (1968). The mean efficiency of equireplicate designs. Biometrika, 55, 251–253.

Pearce, S.C. (1970). The efficiency of block designs in general. Biometrika, 57, 339–346.

Pearce, S.C. (1971). Precision in block experiments. Biometrika, 58, 161-167.

Pearce, S.C., Calinski, T. and Marshall, T.F. dec. (1974). The basic contrasts of an experimental design with special reference to the analysis of the data. Biometrika, 61, 449-460.

Preece, D.A. (1982). Balance and designs: Another terminological tangle. Utilitas Mathematica, 29, C, 85-186.

Puri, P.D. (1975). On efficiency balanced and partially efficiency balanced designs. Unpublished Ph.D. Thesis, IARI, New Delhi.

Puri, P.D. (1984). Patterns and analysis of BIB and G.D. designs having a missing block. Sankhya, B, 46, 44–53.

Puri, P.D., Gupta, V.K. and Nigam, A.K. (1987). A desirable pattern of incomplete block designs for symmetrical parallel line assays. Communicated.

Puri, P.D., and Nigam, A K. (1975a). On patterns of efficiency balanced designs. Journal of the Royal Statistical Society, B, 37, 457–458.

Puri, P.D. and Nigam A.K. (1975b). A note on efficiency balanced designs. Sankhya, B, 37, 457-460.

Puri, P.D. and Nigam, A.K. (1976). Balanced factorial experiments–I. Communications in Statistics, A5(7), 599–619.

Puri, P.D. and Nigam, A.K. (1977a). Partially efficiency balanced designs. Communications in Statistics, A6 (8), 753-771.

Puri, P.D. and Nigam A.K. (1977b). Balanced block designs. Communications in Statistics, A6 (12), 1171–1179.

Puri, P.D. and Nigam, A.K. (1978). Balanced factorial experiments–II. Communications in Statistics, A7 (6), 591–605.

Puri, P.D., Nigam, A.K. and Narain, P. (1979). Supplemented block designs. Sankhya, B, 39, 189–195.

Puri, P.D. and Nigam, A.K. (1983). Merging of treatments in block designs. Sankhya, B, 45, 50–59.

Raghavarao, D. (1962). On balanced unequal block designs. Biometrika, 49, 561–562.

Raghavarao, D. (1966). Duals of partially balanced incomplete block designs and some non-existence theorems. Annals of Mathematical Statistics, 37, 1048–1052.

Raghavarao, D. (1971). Construction and combinatorial problems in design of experiments. John Wiley and Sons. Inc., New York,

Raghavarao, D. and Chandrasekhararao, K. (1964). Cubic designs. Annals of Mathematical Statistics, 35, 389–397.

Ramakrishnan, C.S. (1956). On the dual of a PBIB design and a new class of designs with two replications. Sankhya, 17, 133–142.

Rao, C.R. (1973) Linear statistical inference and its applications. (2nd ed.), John Wiley and Sons, Inc., New York.

Rao, V.R. (1958). A note on balanced designs. Annals of Mathematical Statistics, 29, 290-294.

Rees, D.H. (1966). The analysis of variance of designs with many non-orthogonal classifications Journal of the Royal Statistical Society, B, 28, 110–117.

Roy, J. (1958). On the efficiency factor of a block design. Sankhya, 19, 181–188.

Roy, J. and Laha, R.G. (1956). Classification and analysis of linked block designs. Sankhya, 17, 115–132.

Roy, J. and Laha, R.G. (1957). On partially balanced linked block designs. Annals of Mathematical Statistics, 28, 488-493.

Russell, T.S. and Bradley, R.A. (1958). One way variance in a two-way classification. Biometrika, 45, 111–129.

Saha, G.M. (1975). On construction of balanced ternary designs, Sankhya, B, 37, 220–227

Saha, G.M. (1976). On Calinski's patterns in block designs. Sankhya, B, 38, 383-392.

Saha, G.M. and Dey, A. (1973). On construction and uses of balanced n-ary designs. Annals of the Institute of Statistical Mathematics, 25, 439–445.

Searle, S.R. (1971). Linear models. John Wiley and Sons, Inc. New York.

Searle, S.R. (1975). On reductions in sum of squares when fitting linear models. Technical report M 354, Department of Statistics, Florida State University.

Searle, S.R (1982). Matrix algebra useful for statistics. John Wiley and Sons, Inc., New York.

Seshagiri, A. (1974). Some contribution to design and analysis of bio-assays. Unpublished Ph.D. Thesis, IARI, New Delhi.

Shafiq, M. and Federer, W.T. (1979). Generalized n-ary balanced block designs. Biometrika, 66, 115-123.

Shah, B.V. (1958). On balancing in factorial experiments. Annals of Mathematical Statistics, 29, 766–779.

Shah, B.V. (1960). Balanced factorial experiments. Annals of Mathematical Statistics, 31, 502-514.

Shah, K.R. (1960). Optimality criteria for incomplete block designs. Annals of Mathematical Statistics, 31, 591–594.

Shah, S.M. and Gujarathi, C.C. (1978).O n a locally resistant BIB design of degree one. Sankhya, B, 39, 406-408.

Shah, S.M. and Gujarathi, C.C. (1983). Resistant BIB designs. Sankhya, B, 45, 225-232.

Shrikhande, S.S. (1952). On the duals of some balanced incomplete block designs. Biometrics, 8, 66-72.

Shrikhande, S.S. (1962). On a two parameter family of BIB designs. Sankhya, A, 24, 33-40.

Shrikhande, S.S and Bhagwandas (1965). Duals of incomplete block designs. Journal of the Indian Statistical Association, 3, 30-37.

Singh, M. (1976). Some contributions to the design and analysis of balanced and partially balanced experiments. Unpublished Ph.D. thesis, IARI, New Delhi.

Singh, M. and Dey, A. (1979) On analysis of some augmented block designs. Biometrical Journal, 21, 87-92.

Singh, R. and Raghavarao, D. (1975). Application of linked block designs in successive sampling. Applied Statistics, (Ed.R.P. Gupta). North Holland, 301-309.

Sinha, B.K. (1980). Optimal block designs. Unpublished seminar notes. Stat. Math. Division, ISI, Calcutta (Revised 1982).

Sinha, K., Mathur, S.N. and Nigam, A.K. (1979). Kronecker sum of incomplete block designs. Utilitas Mathematica, 16, 157-164.

Smith, K. (1918). On the standard deviation of adjusted and interpolated values of an observed polynomial function and its constants and the guidance they give towards a proper choice of the distribution of observations. Biometrika, 12, 1-85.

Stevens, W.L. (1948). Statistical analysis of a non-orthogonal tri-factorial experiment. Biometrika, 35, 346-367.

Street, A.P. and Street, D.J. (1987). Combinatorics of experimental designs. Oxford Science Publication, Clarendon Press, Oxford.

Surendran, P.U. (1968). Association matrices and the Kronecker product of the design. Annals of Mathematical Statistics, 39, 676-680.

Takeuchi, K. (1961). On the optimality of certain type of PBIB designs. Rep. Statist. Appl. Res Un Japan Sci. Engrs. 8, 40-145.

Tharthare, S.K. (1963). Right angular designs. Annals of Mathematical Statistics, 34, 1057-1067.

Tharthare, S,K. (1965) Generalized right angular designs. Annals of Mathematical Statistics, 36, 1535-1553.

Tocher, K.D. (1952). The design and analysis of block experiments. Journal of the Royal Statistical Society, B, 14, 45-100.

Tyagi, B.N. (1979) On a class of variance balanced block designs. Journal of the Statistical Planning and Inference, 3, 333-336.

Vartak, M.N. (1955). On an application of Kronecker product of matrices to statistical designs. Annals of Mathematical Statistics, 27, 420-438.

Vartak, M.N. (1959). The non-existence of certain PBIB designs. Annals of Mathematical Statistics, 30, 1051-1062.

Verdooren, L.R. (1966). Anwendung und analyse michtorthogonaler sortenversuche. Vortrag aufdem Internationale symposium fur pflanzensort-enversuche in Budapest.

Wald, A. (1943). On the efficient design of statistical investigations. Annals of Mathematical Statistics, 14, 134–140.

Wilkinson, G.N. (1970). A general recursive procedure for analysis of variance. Biometrika, 57, 19–46.

Williams, E.R. (1975). Efficiency of balanced designs. Biometrika, 62, 686–689.

Williams, E.R. (1976). Resolvable paired comparison designs. Journal of the Royal Statistical Society, B, 38, 171–174.

Williams, E.R., Patterson, H.D. and John, J.A. (1976). Resolvable designs with two replications. Journal of the Royal Statistical Society, B, 38, 296–301.

Wolock, F.W. (1964). Cyclic designs. Ph. D. dissertation, Virgina Polytechnic Institute.

Wynn, H.D. (1977). Convex set of finite population plans. Annals of Statistics, 5, 414–418.

Yamamoto, S., Fuji, Y and Hamada, N. (1965). Composition of some series of association algebra. Journal of Science, Hiroshima University, A 29, 181–215.

Yates, F. (1936). A new method for arranging variety trials involving a large number of varieties. Journal of Agricultural Sciences, Cambridge, 26, 424–455.

Yates F. (1937). The design and analysis of factorial experiments. Imperial Bureau of Soil Science, Technical Communication No. 35.

Yeh, C.M. and Bradley, R.A (1983). Trend free block designs—existence and construction results. Communications in Statistics, Theory and Methods, 12, 1–24.

Yeh, C.M, Bradley, R.A. and Notz. W.I. (1985). Nearly trend free block designs. Journal of the American Statistical Association, 80, 985–992.

Youden, W.G. (1951). Linked blocks: a new class of incomplete block designs (abstract). Biometrics, 7, 124.

Zelen, M. (1954). Analysis for some partially balanced incomplete block designs having a missing block. Biometrics, 10, 273–281.

Zelen, M. (1958). The use of group divisible designs for confounded asymmetrical factorial arrangements. Annals of Mathematical Statistics, 29, 22–40.

Index